THE QUESTION CONCERNING
TECHNOLOGY IN CHINA

THE QUESTION CONCERNING
TECHNOLOGY IN CHINA
An Essay in Cosmotechnics

YUK HUI

URBANOMIC

When I hear modern people complain of being lonely then I know what has happened. They have lost the cosmos.

D.H. Lawrence, *Apocalypse*

If communism in China should come to rule, one can assume that only in this way will China become 'free' for technology. What is this process?

Martin Heidegger, *GA97 Anmerkungen I-V*

For Bernard

Published in 2016 by
URBANOMIC MEDIA LTD,
THE OLD LEMONADE FACTORY,
WINDSOR QUARRY,
FALMOUTH TR11 3EX,
UNITED KINGDOM

Second edition 2018
Third corrected edition 2022

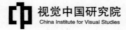
视觉中国研究院
China Institute for Visual Studies

This book is supported by the China Institute for Visual Studies,
China Academy of Art.

BRITISH LIBRARY CATALOGUING-IN-PUBLICATION DATA

A full catalogue record of this book is available
from the British Library

ISBN 978-0-9954550-0-9

Distributed by The MIT Press, Cambridge, Massachusetts
and London, England

Type by Norm, Zurich
Printed and bound in the UK by
TJ Books Limited

www.urbanomic.com

CONTENTS

Part 2. Modernity and Technological Consciousness   199

PREFACE

Quite a few of the notes to which I returned in writing this book date from my teenage years, when I was fascinated both by the cosmogony of Neo-Confucianism and by contemporary astrophysics. I remember how, over several summers, I went regularly every week to the central library in Kowloon with my brother Ben, and brought home piles of books on physics and metaphysics, spending all day reading things that were beyond me and which at the time I didn't know how to use. Luckily, I profited from many discussions with my literature and calligraphy teacher Dr. Lai Kwong Pang, who introduced me to the thought of the New Confucian philosopher Mou Zongsan (1909–1995)—his PhD supervisor at that time. When I started studying Western philosophy, especially contemporary thought, I confronted the great difficulty of integrating it with what I had learned in the past without falling prey to a superficial and exotic comparison. In 2009, an encounter with the work of Keiji Nishitani and Bernard Stiegler on Heidegger suggested to me a way to approach the different philosophical systems from the perspective of the question of time; more recently, while reading the works of anthropologist Philippe Descola and Chinese philosopher Li Sanhu, I began to formulate a concrete question: If one admits that there are multiple natures, is it possible to think of multiple technics, which are different from each other not simply functionally and aesthetically, but also ontologically and cosmologically? This is the principal question of the current work. I propose what I call cosmotechnics as an attempt to open up the question of technology and its history, which for various reasons has been closed down over the last century. This book is only able to present a sketch of such a project, and because it touches upon a long history, expositions of certain subjects of Chinese thought are limited and will have to be supplemented in future.

There are many people to whom I would like to express my gratitude: Prof. Erich Hörl for generously hosting this project and for the discussions; the China Academy of Art for supporting the production of this book, and for discussions with Prof. Gao Shiming, Prof. Guan Huaibin, Prof. Huang Sunquan, Johnson Chang, Lu Ruiyang, Wei Shan, Jiang Jun, Yao Yuchen, Zhang Shunren, Zhou Jing; members of the Pharmakon Philosophy School, Anne Alombert, Sara Baranzoni, Anaïs Nony, Paolo Vignola, Paul-Émile Geoffroy, Michaël Crevoisier, François Corbisier, Axel Andersson, Caroline Stiegler, Elsa Stiegler, Augustin Stiegler, Paul Willemarck (also for his introduction to the work of Rudolf Boehm); colleagues and friends with whom I have had inspiring discussions, Howard Caygill, Scott Lash, Jean-Hugues Barthélémy, Vincent Bontems, Louis Morelle, Louise Piguet, Tristan Garcia, Vincent Normand, Adeena Mey, Regula Bührer, Nathalie Scattolon, Géo Scattolon, Alexandre Monnin, Pieter Lemmens, Armin Beverungen, Marcel Mars, Martina Leeker, Andreas Broeckmann, Holger Fath, Cécile Dupaquier, Jeffrey Shaw.

I would also like to thank Robin Mackay and Damian Veal for their great editorial work, critical comments, and invaluable suggestions. Lastly, I want to thank Bernard Stiegler for the generous discussions and inspirations over the past years.

Yuk Hui
Berlin, Summer 2016

TIMELINE OF THINKERS, EAST AND WEST, DISCUSSED IN THE BOOK

Prehistory
Fu Xi (伏羲)
Nüwa (女媧)
Shennong (神農)
(Yan Di, Lie Shan Xi)

1600–1046 BC: Shang Dynasty

1046–256 BC: Zhou Dynasty

Laozi (老子, –531 BC)	Solon (640–558 BC)
Confucius (孔子, 551–479 BC)	Thales (624–546 BC)
Mozi (墨子, 470–391 BC)	Anaximander (610–546 BC)
Zhuangzi (莊子, 370–287 BC)	Heraclitus (535–475 BC)
Mencius (孟子, 372–289 BC)	Parmenides (515–450 BC)
Xunzi (荀子, 313–238 BC)	Sophocles (497/6–406/5 BC)
	Socrates (470/469–399 BC)
	Plato (428/427–348/347 BC)
	Aristotle (384–322 BC)
	Euclid (300 BC)
221–206 BC: Qin Dynasty	Archimedes (287–212 BC)
	Zeno of Citium (334–262 BC)
	Cleanthes (330–230 BC)
	Chrysippus of Soli (279–206 BC)

206 BC–220 CE: Han Dynasty

Liu An (劉安, 179–122 BC)	Cicero (106–43 BC)
Dong Zhongshu	Seneca (1–65 CE)
(董仲舒, 179–104 BC)	Claudius Ptolemy (100–170)
Sima Qian (司馬遷, 145–90 BC)	Marcus Aurelius (121–180)
Zheng Xuan (鄭玄, 127–200 CE)	

220–589 CE: Six Dynasties

Three Kingdoms (220–265 CE)	
Jin Dynasty (265–420 CE)	
Northern and Southern Dynasties	Pappus of Alexandria (290–350)
(386–589 CE)	Diogenes Laërtius (3rd century)
Wang Bi (王弼, 226–249)	Augustine (354–430)
Guo Xiang (郭象, 252–312)	Boethius (480–524)

581–618: Sui Dynasty ●

618–907: Tang Dynasty ●
Han Yu (韓愈, 768–824)
Liu Zong Yuan
(柳宗元, 773–819)
Hongren (弘忍, 601–685)
Shenxiu (神秀, 606–706)
Heineung (慧能, 638–713)

907–960: Five Dynasties ●

960–1279: Song Dynasty ●
Northern Song (960–1127)
Southern Song (1127–1279)
Zhou Dunyi (周敦頤, 1017–1073)   Adelard of Bath (1080–1152)
Zhang Zai (張載, 1020–1077)   Thomas Aquinas (1225–1274)
Cheng Hao (程顥, 1032–1085)
Chen Yi (程頤, 1033–1107)
Shao Yung (邵雍, 1011–1077)
Zhu Xi (朱熹, 1130–1200)
Dōgen Zenji (道元禅師, 1200–1253)

1206–1368: Yuan Dynasty ●

1368–1644: Ming Dynasty ●
Wang Yangming   Nicholas of Cusa (1401–1464)
(王陽明, 1472–1529)   Bartolomeo Zamberti (1473–1543)
Song Yingxing   Nicolaus Copernicus (1473–1543)
(宋應星, 1587–1666)   Tycho Brahe (1546–1601)
                                      Francisco Suárez (1548–1617)
                                      Galileo Galilei (1564–1642)
                                      Johannes Kepler (1571–1630)
                                      René Descartes (1596–1650)

1644–1911: Qing Dynasty ●
Wang Fuzhi (王夫之, 1619–1692)   Baruch Spinoza (1632–1677)
Dai Zhen (戴震, 1724–1777)   Isaac Newton (1642–1727)
Duan Yucai (段玉裁, 1735–1815)   Gottfried Wilhelm Leibniz (1646–1716)
Zhang Xuecheng (章學誠, 1738–1801)   Immanuel Kant (1724–1804)
Gong Zizhen (龔自珍, 1792–1841)   Georg Wilhelm Friedrich Hegel
Wei Yuen (魏源, 1795–1856)   (1770–1831)
Yan Fu (嚴復, 1894–1921)   Friedrich Wilhelm Joseph
Kang YouWei (康有為, 1858–1927)   von Schelling (1775–1854)

Tan Sitong (譚嗣同, 1865–1898)
Wu Zhihui (吳稚暉, 1865–1953)
Nishida Kitarō (西田幾多郎,1870–1945)
Wang Guo Wei (王國維, 1877–1927)

1912–1949: Republic of China ●
Chen Duxiu (陳獨秀, 1879–1942)
Xiong Shili (熊十力, 1885–1968)
Chang Tungsun(張東蓀, 1886–1973)
Carsun Chang (張君勱, 1887–1968)
Ding Wenjiang (丁文江, 1887–1936)
Watsuji Tetsurō (和辻哲郎, 1889–1960)
Hu Shi (胡適, 1891–1962)
Xu Dishan(許地山, 1893–1941)
Liang Shuming (梁漱溟, 1893–1988)
Feng Youlan (馮友蘭, 1895–1990)
Nishitani Keiji (西谷啓治, 1900–1990)
Mou Zongsan (牟宗三, 1909–1995)
Zhang Dainian (張岱年, 1909–2004)
Yu Guang Yuan (於光遠, 1915–2013)
Lao Szekwang (勞思光, 1927–2012)
Li Zehou (李澤厚, 1930–)
Yu Yingshih (余英時, 1930–)
Chen Changshu (陳昌曙, 1932–2011)
Liu Shuhsien(劉述先, 1934–2016)
Tu Weiming (杜維明, 1940–)

Friedrich Hölderlin (1770–1842)
Ernst Christian Kapp
(1808–1896)
Friedrich Wilhelm Nietzsche
(1844–1900)
Edmund Husserl (1859–1938)
Henri Bergson (1859–1892)
Friedrich Dessauer (1881–1963)
Sigmund Freud (1886–1939)
Martin Heidegger (1889–1976)
Herbert Marcuse (1898–1979)
André Leroi-Gourhan
(1911–1986)
Jacques Ellul (1912–1994)
Jean-Pierre Vernant
(1914–2007)
Gilbert Simondon (1924–1989)
Jean-François Lyotard
(1924–1998)
Jürgen Habermas (1929–)
Jacques Derrida (1930–2004)
Alain Badiou (1937–)
Peter Sloterdijk (1947–)
Bernard Stiegler (1952–2020)

INTRODUCTION

In 1953 Martin Heidegger published his famous lecture 'Die Frage nach der Technik',[1] in which he announced that the essence of modern technology is nothing technological, but rather enframing (*Ge-stell*)—a transformation of the relation between man and the world such that every being is reduced to the status of 'standing-reserve' or 'stock' (*Bestand*), something that can be measured, calculated, and exploited. Heidegger's critique of modern technology opened up a new awareness of technological power, which had already been interrogated by fellow German writers such as Ernst Jünger and Oswald Spengler. Heidegger's writings following 'the turn' (*die Kehre*) in his thought (usually dated around 1930), and this text in particular, portray the shift from *technē* as *poiesis* or *bringing forth* (*Hervorbringen*) to technology as *Gestell*, seen as a necessary consequence of Western metaphysics, and a destiny which demands a new form of thinking: the thinking of the question of the truth of Being.

Heidegger's critique found a receptive audience among Eastern thinkers[2]—most notably in the teachings of the Kyoto School, as well as in the Daoist critique of technical rationality, which identifies Heidegger's *Gelassenheit* with the classical Daoist concept of *wu wei* or 'non-action'. This receptivity is understandable for several reasons. Firstly, Heidegger's pronouncements regarding the power and dangers of modern technology seemed to have been substantiated by the devastations of war, industrialisation, and mass consumerism,

---

1.   M. Heidegger, 'The Question Concerning Technology', in *The Question Concerning Technology and Other Essays*, tr. W. Lovitt (New York and London: Garland Publishing, 1977), 3–35. The original title of the Bremen lecture (1949) is 'Das Gestell'.

2.   In this book, by 'East', I generally mean East Asia (China, Japan, Korea, etc., countries that were influenced by Confucianism, Buddhism, and, to some degree, Daoism).

leading to interpretations of his thought as a kind of existentialist humanism, as in the mid-century writings of Jean-Paul Sartre. Such interpretations resonated deeply with the anxieties and sense of alienation aroused by the rapid industrial and technological transformations in modern China. Secondly, Heidegger's meditations echoed Spengler's claim about the decline of Western civilisation, though in a more profound key—meaning that they could be taken up as a pretext for the affirmation of 'Eastern' values.

Such an affirmation, however, engenders an ambiguous and problematic understanding of the question of technics and technology and—with the arguable exception of postcolonial theories—has prevented the emergence of any truly original thinking on the subject in the East. For it implies a tacit acceptance that there is only one kind of technics and technology,[3] in the sense that the latter are deemed to be anthropologically universal, that they have the same functions across cultures, and hence must be explained in the same terms. Heidegger himself was no exception to the tendency to understand both technology and science as 'international', in contrast to thinking which is not 'international', but unique and 'homely'. In the recently published *Black Notebooks*, Heidegger wrote:

> The 'sciences', like technology and like the technical schools (*Techniken*), are necessarily international. An international thinking does not exist, only the universal thinking, coming from one source.

3.   I make a distinction between the use of the words technics, *technē*, and technology: *technics* refers to the general category of all forms of making and practice; *technē* refers to the Greek conception of it, which Heidegger understands as *poiesis* or bringing forth; and *technology* refers to a radical turn which took place during European modernity, and developed in the direction of ever-increasing automation, leading consequently to what Heidegger calls the *Gestell*.

However, if it is to remain close to the origin, it requires a fateful [*geschicklich*] dwelling in a unique home [*Heimat*] and the unique people [*Volk*], so that it is not the folkish purpose of thinking and the mere 'expression' of people [*des Volkes*]—; the respective only fateful [*geschicklich*] home [*Heimattum*] of the down-to-earthness is the rooting, which alone can enable growth into the universal.[4]

This statement demands further analysis: firstly, the relation between thinking and technics in Heidegger's own thought needs to be elucidated (see §7 and §8, below), and secondly, the problematic of the 'homecoming' of philosophy as a turning against technology needs to be examined. However, it is clear here that Heidegger sees technology as something detachable from its cultural source, already 'international', and which therefore has to be overcome by 'thinking'.

In the same *Black Notebook*, Heidegger commented on technological development in China, anticipating the victory of the Communist Party,[5] in a remark that seems to hint at the failure to address the question concerning technology in China in the decades that would follow the Party's rise to power:

---

4.   '»Wissenschaften« sind, wie die Technik und als Techniken, notwendig international. Ein internationales Denken gibt es nicht, sondern nur das im Einen Einzigen entspringende universale Denken. Dieses aber ist, um nahe am Ursprung bleiben zu können, notwendig ein geschickliches Wohnen in einziger Heimat und einzigem Volk, dergestalt, daß nicht dieses der völkische Zweck des Denkens und dieses nur »Ausdruck« des Volkes—; das jeweilig einzige geschickliche Heimattum der Bodenständigkeit ist die Verwurzelung, die allein das Wachstum in das Universale gewährt.' M. Heidegger, GA 97 *Anmerkungen I-V (Schwarze Hefte 1942–1948)* (Frankfurt Am Main: Vittorio Klostermann, 2015), 59-60, 'Denken und Dichten'.

5.   GA 97 was written between 1942 and 1948; the Chinese Communist Party came to power in 1949.

If communism in China should come to rule, then one can
assume that only in this way does China become 'free' for technology. What is this process?[6]

What does becoming 'free' for technology mean here, if not
to fall prey to an inability to reflect upon it and to transform
it? And indeed, a lack of reflection upon the question of
technology in the East has prevented the emergence of any
genuine critique originating from its own cultures: something
truly symptomatic of a detachment between thinking and
technology similar to that which Heidegger described during
the 1940s in Europe. And yet if China, in addressing this question, relies on Heidegger's fundamentally Occidental analysis
of the history of technics, we will reach an impasse—and
this, unfortunately, is where we stand today. So what is the
question concerning technology for non-European cultures
prior to modernization? Is it the same question as that of the
West prior to modernization, the question of Greek *technē*?
Furthermore, if Heidegger was able to retrieve the question
of Being from the *Seinsvergessenheit* of Western metaphysics, and if today Bernard Stiegler can retrieve the question
of time from the long *oubli de la technique* in Western philosophy, what might Non-Europeans aspire to? If these questions are not even posed, then Philosophy of Technology in
China will continue to be entirely dependent upon the
work of German philosophers such as Heidegger, Ernst Kapp,
Friedrich Dessauer, Herbert Marcuse, and Jürgen Habermas,
American thinkers such as Carl Mitcham, Don Ihde, and Albert

---

6.   'Wenn der Kommunismus in China an die Herrschaft kommen sollte,
steht zu vermuten, daß erst auf diesem Wege China für die Technik »frei«
wird. Was liegt in diesem Vorgang?' Ibid., 441.

Borgmann, and French thinkers such as Jacques Ellul, Gilbert Simondon, and Bernard Stiegler. It seems incapable of moving forward—or even backward.

I believe that there is an urgent need to envision and develop a philosophy of technology in China, for both historical and political reasons. China has modernised itself over the past century in order to 'outstrip the UK and catch up with the US' (超英趕美, a slogan proposed by Mao Zedong in 1957); now it seems to be at a turning point, its modernisation having reached a level that allows China to situate itself among the great powers. But at the same time, there is a general sentiment that China cannot continue with this blind modernisation. The great acceleration that has taken place in recent decades has also led to various forms of destruction, cultural, environmental, social, and political. We are now, so geologists tell us, living in a new epoch—that of the Anthropocene—which began roughly in the eighteenth century with the Industrial Revolution. Surviving the Anthropocene will demand reflection upon—and transformation of—the practices inherited from the modern, in order to overcome modernity itself. The reconstruction of the question of technology in China outlined here also pertains to this task, aiming to unfold the concept of technics in its plurality, and to act as an antidote to the modernisation programme by reopening a truly global history of the world. The book is an attempt both to respond to Heidegger's concept of technics, and to sketch out a possible way to construct a properly *Chinese* philosophy of technology.

§1. THE BECOMING OF PROMETHEUS

Is there technological thought in China? At first glance, this is a question that can be easily dismissed, for what culture doesn't have technics? Certainly, technics has existed in China for many centuries, if we understand the concept to denote skills for making artificial products. But responding to this question more fully will require a deeper appreciation of what is at stake in the question of technics.

In the evolution of human as *homo faber*, the moment of the liberation of the hands also marks the beginning of systematic and transmissible practices of making. They emerge firstly from the need for survival, to make fire, to hunt, to build dwellings; later, as certain skills are gradually mastered so as to improve living conditions, more sophisticated technics can be developed. As French anthropologist and palaeontologist André Leroi-Gourhan has argued, at the moment of the liberation of the hands, a long history of evolution opened up, by way of the exteriorisation of organs and memory and the interiorisation of prostheses.[7] Now, within this universal technical tendency, we observe a diversification of artefacts across different cultures. This diversification is caused by cultural specificities, but also reinforces them, in a kind of feedback loop. Leroi-Gourhan calls these specificities 'technical facts'.[8] While a technical *tendency* is necessary, technical *facts* are accidental: as Leroi-Gourhan writes, they result from the 'encounter of the tendency and thousands of coincidences of the milieu':[9] while the invention of the wheel is a technical tendency, whether or not wheels will have spokes is a matter of technical fact. The early days of the science of making are dominated by the technical tendency, meaning that what reveals itself in human

---

7.   A. Leroi-Gourhan, *Gesture and Speech* (Cambridge, MA and London: MIT Press, 1993).

8.   A. Leroi-Gourhan, *Milieu et Technique* (Paris: Albin Michel, 1973), 336–40; *L'homme et la Matière* (Paris: Albin Michel, 1973), 27–35.

9.   Leroi-Gourhan, *L'homme et la matière*, 27.

activities—for example in the invention of primitive wheels and the use of flint—are optimal natural efficiencies. It is only later on that cultural specificities or technical facts begin to impose themselves more distinctly.[10]

Leroi-Gourhan's distinction between technical tendency and technical fact thus seeks to provide an explanation for the similarities and differences between technical inventions across different cultures. It sets out from a universal understanding of the process of *hominisation* characterised by the technical tendency of invention, as well as the extension of human organs through technical apparatuses. But how effective is this model in explaining the diversification of technologies throughout the world, and the different pace at which invention proceeds in different cultures? It is in light of these questions that I hope to bring into the discussion the dimensions of cosmology and metaphysics, which Leroi-Gourhan himself rarely discussed.

Here is my hypothesis, one which may appear rather surprising to some readers: *in China, technics in the sense we understand it today—or at least as it is defined by certain European philosophers—never existed*. There is a general misconception that all technics are equal, that all skills and artificial products coming from all cultures can be reduced to one thing called 'technology'. And indeed, it is almost impossible to deny that technics can be understood as the extension of the body or the exteriorisation of memory. Yet they may not be *perceived* or reflected upon in the same way in different cultures.

To put it differently, technics as a general human activity has been present on earth since the time of the Australanthropos; but the philosophical concept of technics cannot be

---

10.   Ibid.

assumed to be universal. The technics we refer to here is one that is the subject of philosophy, meaning that it is rendered visible through the birth of philosophy. Understood as such, as a philosophical category, technics is also subject to the history of philosophy, and is defined by particular interrogative perspectives. What we mean by 'philosophy of technology' in this book is not exactly what in Germany is known as *Technikphilosophie*, associated with figures such as Ernst Kapp and Friedrich Dessauer. Rather, it appears with the birth of Hellenic philosophy, and constitutes one of philosophy's core inquiries. And technics thus understood, as an ontological category, I will argue, must be interrogated in relation to a larger configuration, a 'cosmology' proper to the culture from which it emerged.

We know that the birth of philosophy in ancient Greece, as exhibited in the thinking of Thales and Anaximander, was a process of rationalisation, marking a gradual separation between myth and philosophy. Mythology is the source and the essential component of European philosophy, which distanced itself from mythology by naturalizing the divine and integrating it as a supplement to rationality. A rationalist may well argue that any recourse to mythology is a regression, and that philosophy has been able to completely free itself from its mythological origins. Yet I doubt that such a philosophy exists, or ever will. We know that this opposition between *mythos* and *logos* was explicit in the Athenian Academy: Aristotle was very critical of the 'theologians' of the school of Hesiod, and Plato before him argued relentlessly against myth. Through the mouth of Socrates in the *Phaedo* (61a), he says that *mythos* is not his concern but rather the affair of the poets (portrayed as liars in the *Republic*). And yet, as Jean-Pierre Vernant has clearly shown, Plato 'grants an important place in his writings to myth as a means of expressing both those things that lie beyond and those that fall short of strictly philosophical language'.[11]

Philosophy is not the language of blind causal necessity, but rather that which at once allows the latter to be spoken, and goes beyond it. The dialectical movement between rationality and myth constitutes the dynamic of philosophy, without which there would be only positive sciences. The Romantics and German Idealists, writing toward the end of the eighteenth century, were aware of this problematic relationship between philosophy and myth. Thus we read in 'The Oldest System–Programme of German Idealism'—published anonymously in 1797, but whose authors are suspected to be, or at least to be associated with, the three friends from the Tübingen Stift, Hölderlin, Hegel, and Schelling—that 'mythology must become philosophical, and the people rational, and philosophy must become mythological in order to make philosophers sensuous. Then eternal unity reigns among us'.[12] Not coincidentally, this insight came at a moment of renewal of philosophical interest in Greek tragedy, chiefly through the works of these three highly influential friends. The implication here is that, in Europe, philosophy's attempt to separate itself from mythology is precisely conditioned by mythology, meaning that mythology reveals the germinal form of such a mode of philosophising. Every demythologisation is accompanied by a remythologisation, since philosophy is conditioned by an origin from which it can never fully detach itself. Accordingly, in order to interrogate what is at stake in the question of technology, we should turn to

---

11.   J.P. Vernant, *Myth and Society in Ancient Greece*, tr. J. Lloyd (New York: Zone, 1990), 210–11.

12.   'The "Oldest System–Programme of German Idealism"', tr. E. Förster, *European Journal of Philosophy* 3 (1995), 199–200.

the predominant myths of the origins of technology that have been handed down to us, and at once rejected and extended by Western philosophy. The misconception that technics can be considered as some kind of universal remains a huge obstacle to understanding the global technological condition in general, and in particular the challenge it poses to non-European cultures. Without an understanding of this question, we will all remain at a loss, overwhelmed by the homogeneous becoming of modern technology.

Some recent work has attempted to reclaim what it calls 'Prometheanism', decoupling the social critique of capitalism from a denigration of technology and affirming the power of technology to liberate us from the strictures and contradictions of modernity. This doctrine is often identified with, or at least closely related to, the notion of 'accelerationism'.[13] But if such a response to technology and capitalism is applied globally, as if Prometheus were a universal cultural figure, it risks perpetuating a more subtle form of colonialism.

So who is Prometheus, and what does Prometheanism stand for?[14] In Plato's *Protagoras*, the sophist tells the story of the Titan Prometheus, also said to be the creator of human beings, who was asked by Zeus to distribute skills to all living beings. His brother Epimetheus took over the job, but having distributed all the skills, found that he had forgotten to provide

13.   See R. Mackay and A. Avanessian (eds), *#Accelerate: The Accelerationist Reader* (Falmouth and Berlin: Urbanomic/Merve, 2014), especially Ray Brassier's essay 'Prometheanism and its Critics', 469–87.

14.   According to Ulrich von Wilamowitz-Moellendorff, there are two identities of Prometheus: (1) Ionian-Attic Promethos, god of the fire industries, the potter and metalworker honoured in the festival of the Prometheia; and (2) Boeotian-Locrian Prometheus, the Titan whose punishment is part of the great theme of conflict between different generations of the gods. See J.-P. Vernant, *Myth and Thought Among the Greeks* (New York: Zone Books, 2006), 264.

for human beings. In order to compensate for the fault of his brother Epimetheus, Prometheus stole fire from the god Hephaestus and bestowed it upon man.[15] Hesiod told another, slightly different version of the story in his *Theogony*, in which the Titan challenged the omnipotence of Zeus by playing a trick with a sacrificial offering. Zeus expressed his anger by hiding fire and the *means of living* from human beings, in revenge for which Prometheus stole fire. Prometheus received his punishment from Zeus: he was chained to the cliff, and an eagle from Hephaestus came to eat his liver during the daytime and allowed it to grow back at night. The story continues in *Works and Days*, where Zeus, angered by Prometheus's deception (*apatē*) or fraud (*dolos*), revenges himself by visiting evil upon human beings. This evil, or *dolos*, is called Pandora.[16] The figure of Pandora, whose name means 'she who gives everything', is twofold: firstly, she stands for fertility, since in another ancient account, according to Vernant, she has another name, Anesidora, the goddess of the earth;[17] secondly, she stands for idleness and dissipation, since she is a *gastēr*, 'an insatiable belly devouring the *bios* or nourishment that men procure for themselves through their labor'.[18]

It is only in Aeschylus that Prometheus becomes the father of all technics and the master of all crafts (*didasklos technēs pasēs*),[19] whereas before he was the one who stole fire, hiding

15.   Plato, 'Protagoras', tr. S. Lombardo and K. Bell, in J.M. Cooper (ed.) *Complete Works* (Indianapolis, IN: Hackett, 1997), 320c–328d.

16.   Vernant emphasises both acts of Prometheus and Zeus as *dolos*; see Vernant, *Myth and Society*, 185.

17.   Vernant, *Myth and Thought*, 266.

18.   Vernant, *Myth and Society*, 174.

19.   Vernant, *Myth and Thought*, 271.

it in the hollow of a reed.[20] Before Prometheus's invention of technics, human beings were not sensible beings, since they saw without seeing, listened without hearing, and lived in disorder and confusion.[21] In Aeschylus's *Prometheus Bound*, the Titan declares that 'all the *technai* that mortals have, come from Prometheus'. What exactly are these *technai*? It would be difficult to exhaust all possible meanings of the word, but it is worth paying attention to what Prometheus says:

> What's more, for them I invented Number [*arithmon*], wisdom above all others. And the painstaking putting together of Letters: to be their memory of everything, to be their Muses' mother, their handmaid.[22]

In assuming a universal Prometheanism, one assumes that all cultures arise from *technē*, which is originally Greek. But in China we find another mythology concerning the creation of human beings and the origin of technics, one in which there is no Promethean figure. It tells instead of three ancient emperors, who were leaders of ancient tribes (先民): Fuxi (伏羲), Nüwa (女娲) and Shennong (神農).[23] The female goddess Nüwa, who is represented as a half-human, half-snake figure, created human beings from clay.[24] Nüwa's brother, and

---

20. Ibid., 265.

21. Ibid.

22. Aeschylus, *Prometheus Bound*, tr. C. Herrington and J. Scully (New York: Oxford University Press, 1989), 441–506; quoted by D. Roochnik, *Of Art and Wisdom: Plato's Understanding of Techne* (University Park, PA: Pennsylvania State University Press, 1996), 33.

23. There are various accounts of who the three emperors were; the list here is the most commonly used.

24. Concerning the use of clay, different versions of the tale exist: for example,

later husband, is Fuxi, a half-dragon, half-human figure who invented the bagwa (八卦)—the eight trigrams based on a binary structure. Several classical texts document the process whereby Nüwa used five coloured stones to repair the sky in order to stop the water flooding in great expanses and fire blazing out of control.[25] Shennong has quite an ambiguous identity, since he is often associated with two other names, Yan Di (炎帝) and Lie Shan Shi (烈山氏).[26] In this association, Shennong, which literally means 'divine farmer', is also the god of fire, and after his death becomes the god of the kitchen (the character Yan [炎] consists of two repeated instances of the character for fire [火]. It is recognised by historians that it most likely comes from the use of fire in the household, rather than sun worship.)[27] As the name indicates, Shennong also invented agriculture, medicine, and other technics. According to the *Huainanzi*, an ancient Chinese text originating in a series of scholarly debates held at the court of Liu An, King of Huainan (179–122 BC) sometime before 139 BC, he risked poisoning himself by trying hundreds of plants so as to distinguish

according to the *Huainanzi*, the creation of humans was not only the work of Nüwa but a collective work with other gods: 'The Yellow Emperor produced *yin* and *yang*. Shang Pian produced ears and eyes; Sang Lin produced shoulders and arms. Nüwa used these to carry out the seventy transformations.' J. S. Major, S. A. Queen, A. S. Meyer, and H. D. Roth (eds, tr.), *The Huainanzi: A Guide to the Theory and Practice of Government in Early Han China*, *Liu An, King of Huainan* (New York: Columbia University Press, 2010), 17:50. For Chinese see:《淮南子•說林訓》：黃帝生陰陽，上駢生耳目，桑林生臂手：此女媧所以七十化也。

25.　See the *Huainanzi*, chapter 6: 'Surveying Obscurities', 6.7 (《淮南子.覽冥篇》).

26.　Li Gui Min (李桂民), 'The Relation between Shennong, Lie Shan and Yan Di and their Recognition in Antiquity' (神農氏、烈山氏、炎帝的糾葛與遠古傳說的認識問題), *Theory Journal* (理論學刊), 3: 217 (March 2012), 108–12.

27.　Ibid., 109.

what is edible from what is poisonous. The broken sky that Nüwa had to repair resulted from a war between Yan Di's descendant, the god of fire Zhu Rong (祝融) and the god of water Gong Gong (共工).[28] Note that the gods of agriculture and fire came from different systems of mythology, and that, although they are called gods, they are only recognized as such after their deaths—originally, they were leaders of the ancient tribes. Unlike Greek mythology, then, in which the Titan revolted against the gods by bestowing fire and means of subsistence upon human beings, thus raising them above animals, in Chinese mythology there was no such rebellion and no such transcendence granted; this endowment is seen instead as owing to the benevolence of the ancient sages.

In a dialogue with Vernant, French sinologist Jacques Gernet remarked that the radical separation between the world of the gods and the world of man that was necessary for the development of Greek rationality didn't happen in China.[29] Thought of the Greek type did eventually arrive in China, but it arrived too late to exercise any formative influence—the Chinese had already 'naturalised the divine'.[30] In response, Vernant also pointed out that the polar terms characteristic

---

28.   Again, in the Chinese mythologies there are various accounts which differ as to whether Shennong or Nüwa came first, and whether Zhurong is the descendant of Shennong or Huang Di; here we relate the most well-known version.

29.   Vernant, *Myth and Society*, 86.

30.   Gernet also commented elsewhere on the difference between God in Judaism and Christianity and the Heaven in Chinese culture: the former (Jewish and Christian) is the god of pastors, he speaks, commands; while the Chinese heaven does not speak, 'it contents itself to produce the seasons and to act continuously by way of its seasonal influxes'. See J. Gernet, *Chine et Christianisme: action et réaction* (Paris: Gallimard, 1982), 206, cited also by F. Jullien, *Procès ou Création: une introduction à la pensée des lettrés chinois* (Paris: Éditions du Seuil, 1989), 45.

of Greek culture—man/gods, invisible/visible, eternal/mortal, permanent/changing, powerful/powerless, pure/mixed, certain/uncertain—were absent in China, and suggested that this might partially explain why it was the Greeks who invented tragedy.[31]

I do not mean simply to gesture towards the obvious fact that there are different mythologies concerning creation and technics in China, Japan, India, or elsewhere. The point, rather, is that each of these mythologies gives a different origin for technics, corresponding in each case to different relations between the gods, technics, humans, and the cosmos. Apart from some efforts in anthropology to discuss the variation of practices across cultures, these relations have been ignored, or their impact has not been taken into account, in the discourse on technics and technologies. I propose that it is only by tracing different accounts of the genesis of technicity[32] that we can understand what we mean when we speak of different 'forms of life', and thus different relations to technics.

The effort to relativise the concept of technics challenges existing anthropological approaches as well as historical studies, which rest on the comparison of the advancement of either individual technical objects or technical systems (in the sense of Bertrand Gille) in different periods among different cultures.[33]

---

31. Vernant, *Myth and Society*, 98–100.

32. 'Technicity' is a term I borrow from Gilbert Simondon, according to whom technological development should be understood as a genetic process of constant bifurcation that begins during the magical phase of human societies.

33. French historian of technology Bertrand Gille (1920–1980) proposed to analyse the history of technology according to what he calls 'technical systems'. In *Histoire des techniques* (Paris: Gallimard, 1978), 19, Gille defines a 'technical system' as follows: 'All technics are to diverse degrees dependent on one another, and there needs to be a certain coherence between them: this ensemble of the different levels of coherence of all the structures, of all the

Scientific and technical thinking emerges under cosmological conditions that are expressed in the relations between humans and their milieus, which are never static. For this reason I would like to call this conception of technics *cosmotechnics*. One of the most characteristic examples of Chinese cosmotechnics, for example, is Chinese medicine, which uses the same principles and terms found in cosmology, such as *Yin-Yang*, *Wu Xing*, harmony, and so on, to describe the body.

§2. COSMOS, COSMOLOGY, AND COSMOTECHNICS

Here one may ask whether Leroi-Gourhan's analysis concerning technical facts is not already sufficient to explain different technicities. It is true that Leroi-Gourhan brilliantly documented technical tendencies and the diversification of technical facts in his work, documenting different lineages of technical evolution and the influences of the milieu on the fabrication of tools and products. Yet Leroi-Gourhan's research has a limit (even if this also constitutes the strength and singularity of his research), one that seems to stem from his focus upon the individualisation of technical objects so as to construct a technical genealogy and technical hierarchy

---

ensembles and of all the procedures, composes what one can call a technical system.' Technical systems underwent mutation in the face of technological revolutions, for example during the mediaeval period (twelfth and thirteenth centuries), the Renaissance (fifteenth century), and the industrial revolution (eighteenth century). The researchers Yao Dazhi and Per Högselius accused Gille's analysis of being Western-centric, in the sense that Gille used European technical systems as his primary references and, in doing so, ignored Joseph Needham's observation that Chinese technologies seem to have been more advanced than Europe about two thousand years ago. For the debate see Yao Dazhi and P. Högselius, 'Transforming the Narrative of the History of Chinese Technology: East and West in Bertrand Gille's *Histoire des Techniques*', *Acta Baltica Historiae et Philosophiae Scientiarum* 3:1 (Spring 2015), 9–26.

applicable across different cultures. From this perspective, we can understand why he would have deliberately limited himself to an explanation of technical genesis based on the study of the development of tools: as he lamented in the postscript to *L'homme et la matière*, written thirty years after its original publication, most classic ethnographies dedicate their first chapter to technics, only to turn immediately to social and religious aspects for the remainder.[34] In Leroi-Gourhan's work, technics becomes autonomous in the sense that it acts as a 'lens' through which the evolution of the human, civilisation, and culture can be retrieved. However, it is difficult to attribute the singularity of technical facts to the 'milieu' alone, and I do not believe it is possible to avoid the question of *cosmology* and therefore that of *cosmotechnics*.

Allow me to pose this question in the form of a Kantian antinomy: (1) Technics is anthropologically universal, and since it consists in the extension of somatic functions and the externalisation of memory, the differences produced in different cultures can be explained according to the degree to which factual circumstances inflect the technical tendency;[35] (2) Technics is not anthropologically universal; technologies in different cultures are affected by the cosmological understandings of these cultures, and have autonomy only within a certain cosmological setting—technics is always *cosmo*technics. The search for a resolution of this antinomy will be the Ariadne's thread of our inquiry.

I will give a preliminary definition of cosmotechnics here: it means the unification between the cosmic order and the moral order through technical activities (although the term

---

34. Leroi-Gourhan, *L'homme et la matière*, 315.

35. Ibid., 29–35.

cosmic order is itself tautological since the Greek word *kosmos* means order). The concept of cosmotechnics immediately provides us with a conceptual tool with which to overcome the conventional opposition between technics and nature, and to understand the task of philosophy as that of seeking and affirming the organic unity of the two. In the remainder of this Introduction, I will investigate this concept in the work of the twentieth-century philosopher Gilbert Simondon and that of some contemporary anthropologists, notably Tim Ingold.

In the third part of *On the Mode of Existence of Technical Objects* (1958), Simondon sets out a speculative history of technicity, affirming that it is not sufficient just to investigate the technical lineage of objects; it is also necessary to under-stand that it implies 'an *organic* character of thinking and of the mode of being in the world'.[36] According to Simondon, the genesis of technicity begins with a 'magical' phase, in which we find an original unity anterior to the subject/object division. This phase is characterised by the separation and cohesion between ground and figure. Simondon took these terms from Gestalt psychology, where the figure cannot be detached from ground, and it is the ground that gives form, while at the same time form also imposing limitations on the ground. We can conceive the technicity of the magical phase as a field of forces reticulated according to what he calls 'key points' (*pointes clès*), for example high points such as mountains, giant rocks, or old trees. The primitive magical moment, the original mode of cosmotechnics, is bifurcated into technics and religions, in which the latter retain an equilibrium with the former, in the continued effort to obtain unity. Technics and

---

36.   G. Simondon, *Du mode d'existence des objets techniques* (Paris: Aubier, 1958), 213.

religion yield both theoretical and practical parts: in religion, they are known as ethics (theoretical) and dogma (practical); in technics, science and technology. The magical phase is a mode in which there is hardly any distinction between cosmology and cosmotechnics, since cosmology only makes sense here when it is part of everyday practice. There is a separation only during the modern period, since the study of technology and the study of cosmology (as astronomy) are regarded as two different disciplines—an indication of the total detachment of technics from cosmology, and the disappearance of any overt conception of a cosmotechnics. And yet it would *not* be correct to say that there is no cosmotechnics in our time. There certainly is: it is what Philippe Descola calls 'naturalism', meaning the antithesis between culture and nature, which triumphed in the West in the seventeenth century.[37]. In this cosmotechnics, the cosmos is seen as an exploitable standing-reserve, according to what Heidegger calls the world picture (*Weltbild*). Here we should state that for Simondon, there remains some possibility of reinventing cosmotechnics (although he doesn't use the term) for our time. In an interview on mechanology, Simondon talks about the TV antenna, beautifully describing what this convergence (between modern technology and natural geography) should look like. Even though, as far as I am aware, Simondon did not engage further with this subject, it will be our task to take what he meant to say further:

> Look at this TV antenna of a television as it is [...] it is rigid but it
> is oriented; we see that it looks into the distance, and that it can
> receive (signals) from an transmitter far away. For me, it appears

37.   P. Descola, *Beyond Nature and Culture*, tr. J. Lloyd (Chicago and London: Chicago University Press, 2013), 85.

to be more than a symbol; it seems to represent a gesture of sorts, an almost magical power of intentionality, a contemporary form of magic. In this encounter between the highest place and the nodal point, which is the point of transmission of hyperfrequencies, there is a sort of 'co-naturality' between the human network and the natural geography of the region. It has a poetic dimension, as well as a dimension having to do with signification and the encounter between significations.[38]

Retrospectively, we may find that Simondon's proposition is incompatible with the distinction between magic and science made by Lévi-Strauss in *The Savage Mind*, published a few years later (1962). Magic, or rather the 'science of the concrete', according to Lévi-Strauss cannot be reduced to a stage or phase of technical and scientific evolution,[39] whereas for Simondon, as we have seen, the magical phase occupies the first stage of the genesis of technicity. The science of the concrete, according to Lévi-Strauss, is event-driven and sign-oriented, while science is structure-driven and concept-oriented. Thus for Lévi-Strauss there is a discontinuity between the two, but it seems that this discontinuity is only legitimated when one compares a non-European mythical thought with European scientific thought. In Simondon, on the other hand, the magical retains a continuity with the development of science and technology. I would suggest that what Simondon hints at in the third part of *On the Mode of Existence of Technical Objects* is precisely a 'cosmotechnics'. Once we

---

38. G. Simondon, 'Entretien sur la méchanologie', *Revue de synthèse* 130:6, no. 1 (2009), 103–32: 111.

39. C. Lévi-Strauss, *The Savage Mind* (London: Weidenfeld and Nicolson, 1966), 13.

accept the concept of cosmotechnics, instead of maintaining the opposition between the magic/mythical and science and a progression between the two, we will be able to see that the former, characterized as the 'speculative organization and exploitation of the sensible world in sensible terms',[40] is not necessarily a regression in relation to the latter.

Some recent work has suggested that close consideration of non-Western cultures, since it demonstrates a pluralism of ontologies and cosmologies, indicates a way out of the modern predicament. Anthropologists such as Philippe Descola and Eduardo Viveiros de Castro look to Amazonian cultures in order to deconstruct the nature/culture division in Europe. Similarly, philosophers such as François Jullien and Augustin Berque attempt to compare European culture with Chinese and Japanese culture so as to depict a profound pluralism that cannot be easily classified according to simple schemes, and to reinterpret Western attempts to overcome modernity. In his seminal work *Beyond Nature and Culture*, Descola not only suggests that the nature/culture division developed in the Occident is not universal, but also maintains that it is a marginal case. Descola describes four ontologies: namely, naturalism (the nature/culture division), animism, totemism, and analogism. Each of these ontologies inscribes nature in different ways, and in non-modern practices one finds that the nature/culture division that has been taken for granted since European modernity does not hold.[41] Descola cites Social anthropologist Tim Ingold's observation that philosophers have seldom asked, 'What makes humans animals of a particular kind?', their typical preferred question about naturalism being

---

40.   Ibid, 16.

41.   See Descola, *Beyond Nature and Culture*, especially Part III.

'What makes humans different in kind from animals?'[42] This is not only the case among philosophers, as Descola points out; for ethnologists also fall into the dogma of naturalism which insists on the uniqueness of the human being, and the assumption that humans are differentiated from other beings by means of culture.[43] In naturalism, one finds discontinuity in interiority and continuity in physicality; in animism, continuity in interiority and discontinuity in physicality.[44] We reproduce Descola's definitions of the four ontologies below:

Similar interiority, Dissimilar physicality	Animism	Totemism	Similar interiority, Similar physicality
Dissimilar interiority, Similar physicality	Naturalism	Analogism	Dissimilar interiority, Dissimilar physicality

These various ontologies imply different conceptions of nature and different forms of participation; and indeed, as Descola pointed out, the antithesis between nature and culture in naturalism is rejected in other conceptions of 'nature'. What Descola says about nature might also be said of technics, which in Descola's writings is abstracted as 'practice'—a term that avoids the technics/culture division. However, calling it 'practice' may obscure the role of technics; this is the reason we speak of cosmotechnics rather than cosmology.

---

42.   Descola, *Beyond Nature and Culture*, 178.

43.   Ibid., 180.

44.   Ibid., 122.

Although he does not employ a term analogous to 'cosmo-technics', Ingold perceives this point clearly. Drawing on Greg-ory Bateson, Ingold proposes that there is a unity between practices and the environment to which they belong. This leads to his proposal for a sentient ecology,[45] which is medi-ated and operated according to affective relations between human beings and their environments. One example he gives concerning hunter-gatherer society helps to clarify what he means by 'sentient ecology': hunter-gatherers' perception of the environment, he tells us, is embedded in their practices.[46] Ingold points out that the Cree people of northeastern Canada have an explanation for why reindeers are easy to kill: the animals offer themselves voluntarily 'in a spirit of good will or even love towards the hunter'.[47] The encounter between animal and hunter is not simply a question of 'to shoot or not to shoot', but rather one of cosmological and moral necessity:

> At that crucial moment of eye-to-eye contact, the hunter *felt*
> the overwhelming presence of the animal; he felt as if his own
> being were somehow bound up or intermingled with that of
> the animal—a feeling tantamount to love and one that, in the
> domain of human relations, is experienced in sexual intercourse.[48]

Rethinking senses such as vision, hearing, and touch by invok-ing Hans Jonas, James Gibson, and Maurice Merleau-Ponty, Ingold attempts to show that, when we reinvestigate the

---

45.   T. Ingold, *The Perception of the Environment: Essays on Livelihood, Dwelling and Skill* (London: Routledge, 2011), 24.

46.   Ibid., 10.

47.   Ibid., 13.

48.   Ibid., 25.

question of the senses, it is possible to reappropriate this sentient ecology, which is totally ignored in modern techno-logical development. And yet in this conception of human and environment, the relation between environment and cosmol-ogy is not very clear, and this way of analysing living beings with the environment risks reduction to a cybernetic feedback model such as Bateson's, thereby undermining the absolutely overwhelming and contingent role of the cosmos.

Simondon holds a similar view on the relation between human being and the outer world as figure and ground—a functioning model of cosmotechnics, since the ground is limited by the figure, and the figure is empowered by the ground. Owing to their detachment, in religion the ground is no longer limited by the figure, and therefore the unlimited ground is conceived as a godlike power; whereas inversely, in technics, the figure overtakes the ground and leads to the subversion of their relation. Simondon therefore proposes a task for philosophical thinking: to produce a convergence that reaffirms the unity of figure and ground,[49] something that could be understood as the search for a cosmotechnics. For example, in considering Polynesian navigation—the ability to navigate among a thousand islands without any modern apparatus—as a cosmotechnics, we might focus not on this ability as a skill, but rather on the figure-ground relation that prefigures this skill.

The comparison between the work of Ingold and other ethnologists and Simondon indicates two different ways in which the question concerning technology in China might be approached. In the first, we are given a way in which to compre-hend cosmology, which conditions social and political life; while in the second, philosophical thought is reconfigured as a search

---

49.   Simondon, *Du mode d'existence des objets techniques*, 217–18.

for the ground of the figure, whose relation seems to be more and more distanced due to the increasing specialization and division of professions in modern societies. The cosmotechnics of ancient China and the philosophical thought developed throughout its history seem to me to reflect a constant effort to bring about precisely such a unification of ground and figure.

In Chinese cosmology, one finds a sense other than vision, hearing, and touch. It is called *Ganying* (感應), literally meaning 'feeling' and 'response', and is often (as in the work of sinologists such as Marcel Granet and Angus Graham) understood as 'correlative thinking';[50] I prefer to call it *resonance*, following Joseph Needham. It yields a 'moral sentiment' and further, a 'moral obligation' (in social and political terms) which is not solely the product of subjective contemplation, but rather emerges from the resonance between the Heaven and the human, since the Heaven is the ground of the moral.[51] The existence of such a resonance rests upon the presupposition of unification between the human and the Heaven (天人合一), and therefore *Ganying* implies (1) a homogeneity in all beings, and (2) an organicity of the relation between part and part,

---

50. A.C. Graham, *Yin-Yang and the Nature of Correlative Thinking* (Singapore: National University of Singapore, 1986)

51. Concerning the origin of the moral order, it is difficult, for instance, to find an explanation in Henri Bergson's *The Two Sources of Morality and Religion* (tr. A. Audra and C. Brereton [London: Macmillan, 1935]). Bergson distinguishes two kinds of morals: one is a closed morals related to social obligation and habitude, while the other is what he calls an open morals related to 'call of the hero [*appel du héro*]'. In the latter form, one doesn't yield to pressure, but to fascination; according to Bergson these two forms of the moral coexist, and neither exists in pure form. It would certainly be worthwhile to further examine Bergson's concept of the moral and its implications for the Chinese cosmotechnics that I attempt to sketch out here, although it seems to me that Bergson's understanding of the moral is rather limited to the Western tradition, especially the Greeks; in China, the cosmos played a determining role, so that any heroic act could only be an accordance with the Heaven.

and between part and whole.[52] This homogeneity can be found already in *Zhou Yi—Xi Ci* II,[53] where the ancient Bao-xi (another name for Fuxi) created the eight trigrams to reflect the connection of all being through these homogeneities:

> Anciently, when Bao-xi had come to the rule of all under Heaven, looking up, he contemplated the brilliant *forms* exhibited in the sky, and looking down he surveyed the *patterns* shown on the earth. He contemplated the ornamental *appearances* of birds and beasts and the (different) *suitabilities* of the soil. Near at hand, in his own person, he found things for consideration, and the same at a distance, in things in general. On this he devised the eight trigrams, to show fully the attributes of the spirit-like and intelligent (operations working secretly), and to classify the qualities of the myriads of things.[54]

Words such as 'forms', 'patterns', and 'appearances' are essential in understanding the resonances between the Heaven and the human. They imply an attitude towards science in China which (according to the organismic readings offered by authors such as Joseph Needham) differs from that of Greece, since it is resonance that lends authority to rules and laws, whereas for the Greeks laws (*nōmoi*) are closely related

---

52. Huang Junjie (黃俊傑), 東亞儒學史的新視野 [*New Perspectives on the History of Confucianism in East Asia*] (Taipei: Taiwan National University Press, 2015), 267.

53. According to historical documents, there were three versions of the *I Ching* (易經, or *The Book of Changes*) in China, but only one, *Zhou Yi* (周易), has been preserved and circulated. There are seven classic commentaries on the *I Ching*, known as Yi Zhuan (易傳), including the Xi Ci quoted below; together, these ten texts are known as the 'ten wings'.

54. *Xi Ci* II, tr. J. Legge, <http://ctext.org/book-of-changes/xi-ci-xia/ens> [emphasis mine].

to geometry, as Vernant frequently points out. But how is this resonance to be sensed? Confucianism and Daoism both postulate a cosmological 'heart' or 'mind' (examined in §18 below) able to resonate with the external environment (for example in *Luxuriant Dew of the Spring and Autumn Annals*)[55] as well as with other beings (for example in *Mencius*). We will see later how it is this sense that leads to the development of a moral cosmology or moral metaphysics in China, which is expressed in the unification between the Heaven and the human. Importantly for our argument here, in the context of technics such unification is also expressed as the unification of *Qi* (器, literally translated as 'tools') and *Dao* (道, often transliterated as 'tao'). For example, in Confucianism, *Qi* implies a cosmological consciousness of the relations between humans and nature that is demonstrated in rituals and religious ceremonies. As we discuss in Part 1, the Confucian classic *Li Ji* (the *Book of Rituals*) contains a long section entitled *Li Qi* (禮器, 'the vessels of rituals') documenting the importance of technical objects in the fulfilment of the *Li* (禮, 'rituals'), and according to which morality can only be maintained through the proper use of *Li Qi*.

It will be the task of Part 1 to elaborate on this 'correlative thinking' in China, and on the dynamic relation between *Qi* and *Dao*. I believe that the concept of cosmotechnics allows us to trace different technicities, and contributes to opening up the plurality of relations between technics, mythology, and cosmology—and thereby to the embracing of the different relations between the human and technics inherited from different mythologies and cosmologies. Certainly Prometheanism

---

55.   Authorship of this work is attributed to the important Han Confucian Dong Zhongshu (董仲舒, 179–104 BC), who we will discuss below.

is one such relation, but it is highly problematic to take it as a universal. However, I am certainly not proposing to advocate any kind of cultural purity here, or to defend it, as origin, against contamination. Technics has served as a means of communication between different ethnic groups, which immediately calls into question any concept of an absolute origin. In our technological epoch, it is the driving force of globalisation—in the sense both of a converging force acting through space, and a synchronising force in time. Yet a radical alterity will have to be asserted in order to leave room for heterogeneity, and thereby to develop different *epistemes* based on traditional metaphysical categories, a task which opens the way to the veritable question of locality. I use the term *episteme* with reference to Michel Foucault, for whom it denotes a social and scientific structure that functions as a set of criteria of selection, and determines the discourse of truth.[56] In *The Order of Things*, Foucault introduces a periodisation of three *epistemes* in the Occident: Renaissance, Classical, and Modern. Foucault later found that his introduction of the term *episteme* had led to an impasse, and developed a more general concept, namely that of the *dispositif*.[57] The transition from *episteme* to *dispositif* is a strategic move to a more immanent critique, which Foucault was able to apply in a more contemporary analysis; looking back

---

56. M. Foucault, *The Order of Things: An Archeology of the Human Sciences* (New York: Vintage Books, 1994), xxi: 'What I am attempting to bring to light is the epistemological field, the episteme in which knowledge, envisaged apart from all criteria having reference to its rational value or to its objective forms, grounds its positivity and thereby manifests a history which is not that of its growing perfection, but rather that of its conditions of possibility; in this account, what should appear are those configurations within the space of knowledge which have given rise to the diverse forms of empirical science.'

57. M. Foucault, 'Le jeu de Michel Foucault (Entretien sur l'histoire de la sexualité)', in *Dits et Écrits III* (Paris: Gallimard, 1994), 297–329: 301.

during an interview in 1977, around the time of the publication of the *History of Sexuality*, Foucault proposed to define *episteme* as a form of *dispositif*: as that 'strategic *dispositif* which allows the selection, among all possible enunciations, of those that will be acceptable within [...] a field of scientificity of which one can say: this is true or false'.[58] I take the liberty of reformulating the concept of *episteme* here: for me it is a *dispositif* which, in the face of modern technology, may be reinvented on the basis of the traditional metaphysical categories in order to reintroduce a form of life and to reactivate a locality. Such reinventions can be observed, for example, following the social, political, and economic crises that occurred in each epoch in China (and we can surely find examples in other cultures): the decline of the Zhou Dynasty (1122–256 BC), the introduction of Buddhism in China, the country's defeat in the Opium Wars, etc. At these points we observe the reinvention of an *episteme*, which in turn conditions aesthetic, social, and political life. The technical systems that are in the process of forming today, fuelled by digital technologies (for example, 'smart cities', the 'internet of things', social networks, and large-scale automation systems) tend to lead to a homogeneous relation between humanity and technics—that of intensive quantification and control. But this only makes it more important and more urgent for different cultures to reflect on their own history and ontologies in order to adopt digital technologies without being merely synchronized into the homogenous 'global' and 'generic' *episteme*.

The decisive moment in modern Chinese history came with the two Opium Wars in the mid-nineteenth century, in which the Qing dynasty (1644–1912) was comprehensively defeated

---

58.   Ibid.

by the British army, leading to the opening up of China as a quasi-colony for Western forces, and instigating China's modernisation. Lack of technological competence was considered by the Chinese to be one of the major reasons for this defeat. They therefore felt with urgency the need for rapidly modernisation via technological development, in the hope of putting an end to the inequality between China and the Western forces. However, China was not able to absorb Western technology in the way that the dominant Chinese reformists at that time wished, largely due to the ignorance and misunderstanding of technology. For they maintained a belief, which retrospectively seems rather 'Cartesian', that it would be possible to separate Chinese thought—the mind—from technologies understood merely as instruments; that the former, the ground, could remain intact without being affected by the importation and implementation of the technological figure.

On the contrary, technology has ended up subverting any such dualism, and has constituted itself as ground rather than as figure. More than a century and a half has passed since the Opium Wars. China has lived through further catastrophes and crises owing to the change of regimes and all manner of experimental reforms. During this time there have been many reflections on the question of technology and modernisation, and the attempt to maintain a dualism between thinking mind and technological instrument has been revealed as a failure. More seriously, in recent decades any such reflection has been rendered impotent in the face of continuing economic and technological booms. A kind of ecstasy and hype has emerged in its stead, propelling the country into the unknown: all of a sudden, it finds itself as if in the midst of an ocean without being able to see any limit, any destination—the predicament described by Nietzsche in *The Gay Science*, and which remains

a poignant image for describing modern man's troubling situation.[59] In Europe, various concepts such as the 'postmodern'
or 'posthuman' have been invented to name some imaginary
exodus from this situation; but it will not be possible to find the
exit without directly addressing and confronting the question
of technology.

With all of the above questions in mind, this work aims to
open up a new inquiry into modern technology, one that does
not take Prometheanism as its fundamental presupposition.
The work is divided into two parts. Part 1 is intended to be a
systematic and historical survey of 'technological thought' in
China in comparison to its counterpart in Europe. It serves as
a new starting point for understanding what is at stake here,
as well as for reflecting on the urgency of this investigation.
Part 2 is an investigation into the historical-metaphysical
questions of modern technology, and aims to shed new light
on the obscurity in which the question of technology dwells
in China, especially in the Anthropocene.

§3. TECHNOLOGICAL RUPTURE
AND METAPHYSICAL UNITY

As implied by the concept of cosmotechnics outlined above,
the account of technology given here does not limit itself to
the historical, social, and economic levels; we have to move
beyond these levels in order to reconstitute a metaphysical
unity. By 'unity', I do not mean a political or cultural identity, but
a unity between practice and theory, or more precisely a *form
of life* that maintains the coherence (if not necessarily the harmony) of a community. The fragmentation of forms of life in

---

59. F. Nietzsche, *The Gay Science*, tr. J. Nauckhoff (Cambridge: Cambridge
University Press, 2001), 119 (§124).

both European and non-European countries is largely a result of an inconsistency between theory and practice. But in the East this gap is revealed not as a mere disturbance but as the 'deracination' (*Entwurzelung*) described by Heidegger—as a total discontinuity. The transformation of practices brought about by modern technology outstrips the ancient categories that had previously applied. For example, as I discuss in Part 1, the Chinese have no equivalents of the categories that the Greeks called *technē* and *physis*. Hence in China the force of technology dismantles the metaphysical unity of practice and theory, and creates a rupture, which still awaits unification. Of course, this is not something that is only happening in the East. In the West, as Heidegger described, the emergence of the category 'technology' no longer shares the same essence as *technē*. The question concerning technology should ultimately serve as a motivation to take up the question of Being—and, if I might say so, to create a new metaphysics; or, even better, a new cosmotechnics.[60] In our time, this unification or indifference does not present itself as a quest for a ground, but rather exhibits itself as both an original ground (*Urgrund*) and an unground (*Ungrund*): *Ungrund* because it is open to alterities; *Urgrund* as a ground that resists assimilation. Hence the *Urgrund* and the *Ungrund* should be considered as a unity, much like being and nothingness. The quest for unity is properly speaking the *telos* of philosophy, as Hegel maintained in his treatise on Schelling and Fichte.[61]

---

60. Although Heidegger did not explicitly make this claim, in his commentary on Nietzsche he refers to metaphysics as a force of unification that overlooks all beings. However, we have to bear in mind that Heidegger's reading of the history of Western metaphysics is only one possible interpretation: see M. Heidegger, *GA 6.2 Nietzsche Band II* (Frankfurt am Main: Klostermann, 1997), 342–3.

61. G.W.F. Hegel, *The Difference between Fichte's and Schelling's System*

As we shall see, to answer the question concerning technology in China is not to give a detailed history of the economic and social development of technologies—something that historians and sinologists such as Joseph Needham have already done in various brilliant ways—but rather to describe the transformation of the category *Qi* (器) in its relation to *Dao* (道). Let me be more precise on this point. Normally technics and technology are translated in Chinese as *jishu* (技術) and *keji* (科技). The first term means 'technique' or 'skill'; the second is composed of two characters, *ke* meaning 'science' (*ke xue*) and *ji* meaning 'technique' or 'applied science'. The question is not whether these translations adequately render the meanings of the Western words (one has to note that the translations are newly-coined terms), but rather whether they create the illusion that Western technics have an equivalent in the Chinese tradition. Ultimately, the eagerness these Chinese neologisms express to show that 'we also have these terms' obscures the true question of technics. Rather than relying on these potentially confusing neologisms, therefore, I propose to reconstruct the question of technics from the ancient philosophical categories *Qi* and *Dao*, tracing various turning points at which the two were separated, reunified, or even totally disregarded. The relation between *Qi* and *Dao* characterises, properly speaking, the thinking of technics in China, which is also a unification of moral and cosmological thinking in a cosmotechnics. It is in associating *Qi* and *Dao* that the question of technics reaches its metaphysical ground. It is also in entering into this relation that *Qi* participates in moral cosmology, and intervenes in the metaphysical system according to its own evolution. Thus we will show how the relation between *Qi* and *Dao* has varied throughout the history of Chinese thought, following continual attempts to reunify *Dao* and *Qi* (道器合一), each with different nuances and different consequences: *Qi* enlightens *Dao* (器以明道), *Qi* carries *Dao* (器以載道) or *Qi* in the service of *Dao* (器為道用), *Dao* in the service of *Qi* (道為器用), and so on. Below we trace these relations from the era of Confucius and Laozi into contemporary China. Finally, we show how the imposition of a superficial and reductive materialism ended up completely separating *Qi* and *Dao*, an event that may be considered as the breakdown of the traditional system, and may even be termed China's own 'end of metaphysics'—although once again, here we should emphasise that what is called 'meta-physics' in the European language is not equivalent to its usual translation in Chinese, *Xing er Shang Xue* (形而上學), which actually means 'that which is above forms', and is a synonym of *Dao* in the *I Ching*. What Heidegger terms the 'end of metaphysics', then, is by no means the end of *Xing er Shang Xue*—because, for Heidegger, it is the completion of meta-physics that gives us modern technoscience; whereas *Xing er Shang Xue* cannot give rise to modern technology, since firstly it doesn't have the same source as the *metāphysikā*, and secondly, as we will explain in detail below, if we follow New Confucian philosopher Mou Zongsan, Chinese thought has always given priority to the noumenon over the phenomenon, and it is precisely because of this philosophical attitude that a different cosmotechnics developed in China.

It is not my aim, however, to argue that the traditional Chinese metaphysics is sufficient and that we can simply go back to it. On the contrary, I would like to show that, while it is insufficient to simply revive the traditional metaphysics, it is crucial that we *start* from it in order to seek ways other than affirmative Prometheanism or neocolonial critique to think and to challenge global technological hegemony. The ultimate task will be to reinvent the *Dao-Qi* relation by situating it historically, and asking in what way this line of thinking might be fruitful not only in the construction of a new technological thought via China, but also in responding to the current state of technological globalisation.

Inevitably, this task will also have to respond to the haunting dilemma of what is called 'Needham's question': Why didn't modern science and technology emerge in China? In the sixteenth century, Europeans were attracted by China: by its aesthetics and its culture, but also by its advanced technologies. For example, Leibniz was obsessed with Chinese writing, especially by his discovery that the *I Ching* is organised according to precisely the binary system he himself had proposed. He thus believed he had discovered in the Chinese writings an advanced mode of combinatorics. After the sixteenth century, though, science and technology in China were outstripped by the West. According to the dominant view, it is the modernisation of science and technology in Europe during the sixteenth and seventeenth centuries that accounts for this change. Such an explanation is 'accidental' in the sense that it relies on a rupture or an event; but as we shall try to elaborate, there may be another explanation, from the standpoint of metaphysics.

In asking why modern science and technology did not emerge in China, we will discuss the tentative answers given both by Needham himself, and by the Chinese philosophers Feng Youlan (1895–1990) and Mou Zongsan (1909–1995). Mou's answer is the most sophisticated and speculative of the two, and the solution he proposes demands a reunification of two metaphysical systems: one that speculates on the

noumenal world and makes it the core constituent of a moral
metaphysics, and another that tends to limit itself to the level
of phenomena, and in doing so furnishes the terrain for highly
analytical activities. This reading is clearly influenced by Kant,
and indeed Mou frequently employs Kant's vocabulary. Mou
recalls that, when he first read Kant, he was struck by the fact
that what Kant calls the noumenon is at the core of Chinese
philosophy, and that it is the respective focus on noumenon
and phenomenon that marks the difference between Chinese
and European metaphysics.[62] Indulging in speculation on the
noumenon, Chinese philosophy tends to advance the activities of intellectual intuition, but refrains from dealing with the
phenomenal world: it pays attention to the latter only in order to
take it as a stepping stone to reach 'above form'. Mou therefore
argues that in order to revive traditional Chinese thought, an
interface has to be reconstructed between noumenal ontology and phenomenal ontology. This connection cannot come
from anywhere other than the Chinese tradition itself, since
ultimately Mou means it to be a proof that traditional Chinese
thought *can also* develop modern science and technology, and
only needs a new method in order to do so. This sums up the
task of the 'New Confucianism'[63] which developed in Taiwan
and Hong Kong after the Second World War, and which we
discuss in Part 1 (§18). However, Mou's proposal remains an
idealist one, because he considers *Xin* (心, 'heart'), or the
noumenal subject, as the ultimate possibility: according to him,

---

62. Mou Zongsan, *Collected Works 21: Phenomenon and Thing-in-Itself*
(現象與物自身) (Taipei: Linking Books, 2003), 20–30.

63. It is necessary to distinguish *Neo-Confucianism*, a metaphysical
movement that culminated during the Sung and Ming dynasties, from *New
Confucianism*, which is a movement that started in the early twentieth century.

though, through self-negation it can descend so as to become a subject of (phenomenal) knowledge.[64]

Part 2 of the book serves as a critique of Mou's approach, and proposes to go 'back to the technical objects themselves', as an alternative (or better, a supplement) to this idealist tendency.

§4. MODERNITY, MODERNISATION, AND TECHNICITY

In attempting to think through Mou's proposition of an interface between Chinese and Western thought, while avoiding his idealism, Part 2 finds that what is central here is the relation between technics and time. Here I turn to Bernard Stiegler's reformulation of the history of Western philosophy according to the question of technicity in *Technics and Time*. But time has never been a *real* question for Chinese philosophy; as sinologists Marcel Granet and François Jullien have stated clearly, the Chinese never really elaborated on the question of time.[65] This therefore opens up the possibility, in the wake of Stiegler's work, of an investigation into the *relation* between technics and time in China.

Based on the work of Leroi-Gourhan, Husserl, and Heidegger, Stiegler attempts to put an end to a modernity characterised by *technological unconsciousness*. Technological consciousness is the consciousness of time, of one's finitude; but also of the relation between this finitude and technicity. Stiegler convincingly shows how, from Plato on, the relation between technics and anamnesis is already well established, and stands at the centre of the economy of the soul.

---

64. Mou himself claims that he is not an idealist, since *xin* is not the mind; it is more than the mind, and offers more possibilities.

65. F. Jullien, *Du Temps* (Paris: Biblio Essais, 2012).

After reincarnation, the soul forgets the knowledge of truth that it has acquired in the past life, and the search for truth is fundamentally an act of remembering or recollection. Socrates famously demonstrates this in the *Meno*, where the young slave, with the aid of technical tools (drawing in the sand), is able to solve geometrical problems of which he has no prior knowledge at all.

The economy of the soul in the East, though, has little in common with such an anamnesic conception of time. We must say that, even though the calendrical devices of the cultures resemble each other, in these technical objects we find not only different technical lineages, but also different interpretations of time, which configure the function and perception of these technical objects in everyday life. This is largely the result of the influence of Daoism and Buddhism, which combined with Confucianism to produce what Mou Zongsan calls the 'synthetic approach to comprehending reason [綜合的盡理之精神]' in contrast to occidental culture's 'analytic approach to comprehending reason [分解的盡理之精神]'.[66] In the noumenal experience implied by the former, there simply *is no time*; or more precisely, time and historicity do not occur as questions. In Heidegger, historicity is the hermeneutics conditioned by the finitude of Dasein and technics, which infinitises Dasein's retentional finitude by passing exteriorised memory from generation to generation. Mou appreciated Heidegger's critique of Kant in *Kant and the Problem of Metaphysics*, in which Heidegger radicalised the transcendental imagination, making it a question of time. However Mou also sees Heidegger's analysis of finitude as a limitation, since for Mou, *xin* qua noumenal subject

---

66. Mou Zongsan, *Collected Works 9: Philosophy of History* (歷史哲學) (Taipei: Linking Books, 2003), 192–200.

is that which can indeed 'infinitise'. Mou did not formulate any material relation between technics and the *xin*, since he largely disregarded the question of technics, which, for him, is only one of the possibilities of the self-negation of the *Liangzhi* (heart/mind) (良知的自我坎陷). It is to this lack of reflection on the question of technics, I speculate, that we can attribute the failure of New Confucianism to respond to the problem of modernisation and the question of historicity; however, it is possible and necessary to transform this lack into a *positive* concept, a task akin to that undertaken by Jean-François Lyotard, as we shall examine below.

This disregarding of time and lack of any discourse on historicity in Chinese metaphysics was noted by Keiji Nishitani (1900–1990), a Japanese philosopher of the Kyoto School who studied under Heidegger in Freiburg during the 1930s. For Nishitani, Eastern philosophy did not take the concept of time seriously, and hence was unable to account for concepts such as historicity—that is, the ability to think as a 'historical being'. This question is indeed a most Heideggerian one: in the second division of *Being and Time*, the philosopher discussed the relation between individual time and the relation to *Geschichtlichkeit* (historicity). But in Nishitani's attempt to think East and West together, two problems arise, and present a dilemma. Firstly, for the Japanese philosopher, technology opens a path towards 'nihility', as do the works of Nietzsche and Heidegger; but in the Buddhism espoused by Nishitani, *śûnyatā* (emptiness) aims to transcend nihility; and in such transcendence, time loses all meaning.[67] Secondly, *Geschichtlichkeit* and, further, *Weltgeschichtlichkeit* (world

---

67. K. Nishitani, *Religion and Nothingness* (Berkeley, CA: University of California Press, 1982).

historicity) are not possible without a retentional system—which, as Stiegler shows in the third volume of *Technics and Time*, is also technics.[68] This means that it is not possible to be conscious of the relation between Dasein and historicity without being conscious of the relation between Dasein and technicity—that is to say, historical consciousness demands technological consciousness.

As I argue in Part 2, modernity functions according to a technological unconsciousness, which consists of a forgetting of one's own limits, as described by Nietzsche in *The Gay Science*: 'the poor bird that has felt free and now strikes against the walls of this cage! Woe, when homesickness for the land overcomes you, as if there had been more freedom there—and there is no more "land".'[69] This predicament arises precisely from a lack of awareness of the instruments at hand, their limits and their dangers. Modernity ends with the rise of a technological consciousness, meaning both the consciousness of the power of technology and the consciousness of the technological condition of the human. In order to tackle the questions raised by Nishitani and Mou Zongsan, it is necessary to articulate the question of time and history with that of technics, so as to open up a new terrain and to explore a thinking that bridges noumenal ontology and phenomenal ontology.

But in demanding that a Chinese philosophy of technology adopt this post-Heideggerian (Stieglerian) viewpoint, aren't we in danger of simply imposing a Western point of view once again? Not necessarily, since what is more fundamental today is to seek a new conception of world history

---

68. B. Stiegler, *Technics and Time 3: Cinematic Time and the Question of Malaise*, tr. S. Barker (Stanford, CA: Stanford University Press, 2010).

69. Nietzsche, *The Gay Science*, 119.

and a cosmotechnical thinking that will give us a new way of being with technical objects and systems. Far from simply renouncing the analyses of Mou and Nishitani and replacing it with Stiegler's, we therefore pose the following question: Rather than absorbing technics into either of their ontologies, is it possible to understand technics as a *medium* for the two ontologies? For Nishitani, the question was: Can absolute nothingness appropriate modernity and hence construct a new world history that is not limited by Western modernity? For Mou: Can Chinese thinking absorb modern science and technology through a reconfiguration of its own thinking that already lies within the possibilities of the latter? Nishitani's answer leads to a proposal for a total war as a strategy to overcome modernity, something that was taken up as the slogan of the Kyoto school philosophers prior to the Second World War. This is what I term a metaphysical fascism, which arises from a misdiagnosis of the question of modernity, and is something we must avoid at all costs. Mou's answer was affirmative and positive even if, as we will see in Part 1, it was widely questioned by Chinese intellectuals. It seems to me that both Mou and Nishitani (as well as their schools and the epochs in which they lived) failed to overcome modernity largely because they didn't take the question of technology seriously enough. However, we still have to pass through their work in order to clarify these problems. One point that can be stated clearly here is that, in order to heal the rupture of the metaphysical system introduced by modern technology, we cannot rely on any speculative idealist thinking. Instead, it is necessary to take the materiality of technics (as *ergon*) into account. This is not a materialism in the classical sense, but one that pushes the possibility of matter to its limits.

This question is at once speculative and political. In 1986, Jean-François Lyotard, on the invitation of Bernard Stiegler, gave a seminar at IRCAM, at the Centre Pompidou in Paris, later published under the title '*Logos* and *Techne*, or Telegraphy'.[70] In the seminar Lyotard asked whether it is possible that, rather than being retentional devices, the new technologies might open up a new possibility of thinking what the thirteenth-century Japanese Zen Buddhist Dōgen calls the 'clear mirror [明鏡]'. Lyotard's question resonates with the analyses of Mou and Nishitani, since the 'clear mirror' fundamentally constitutes the heart of the metaphysical systems of the East. Towards the end of the talk, Lyotard concludes as follows:

> The whole question is this: is the passage possible, will it be possible with, or allowed by, the new mode of inscription and memoration that characterizes the new technologies? Do they not impose syntheses, and syntheses conceived still more intimately in the soul than any earlier technology has done? But by that very fact, do they not also help to refine our anamnesic resistance? I'll stop on this vague hope, which is too dialectical to take seriously. All this remains to be thought out, tried out.[71]

Why did Lyotard, having made this proposal, retreat from it, suggesting that it was too vague and too dialectical to be taken seriously? Lyotard approached the question from the opposite direction to Mou Zongsan and Keiji Nishitani: he was looking for a passage from West to East. However, Lyotard's limited knowledge of the East did not allow him to go further, into the question of world historicity.

---

70.   J.-F. Lyotard, *The Inhuman: Reflections on Time*, tr. G. Bennington and R. Bowlby (Cambridge: Polity, 1991).

71.   Ibid., 57.

Along with many others of his time, notably Bruno Latour, Lyotard is a representative of the second attempt of European intellectuals to overcome modernity. The first attempt was around the time of the First World War, when intellectuals were conscious of the decline of the West and the crisis that was presenting itself in the domains of culture (Oswald Spengler), science (Edmund Husserl), mathematics (Hermann Weyl), physics (Albert Einstein), and mechanics (Richard von Mises). In parallel, East Asia saw the first generation of New Confucians (Xiong Shili, the teacher of Mou Zongsan, and Liang Shuming) and intellectuals such as Liang Qichao and Zhang Junmai; the very much germanised Kyoto school; and then the second generation of New Confucians in the 1970s—all of whom attempted to broach the same questions. However, like the first generation of New Confucians, they remained insensitive to their idealist approach towards modernisation, and didn't give the question of technology the properly philosophical status that it deserves. In Europe we are now witnessing a third attempt, with anthropologists such as Descola and Latour, who seek to use the event of the Anthropocene as an opportunity to overcome modernity in order to open up an ontological pluralism. In parallel, in Asia, we also see the efforts of scholars who are seeking ways to understand modernity without relying on European discourse—notably the Inter-Asia School initiated by Johnson Chang and others.[72]

§5. WHAT IS THE 'ONTOLOGICAL TURN' FOR?

For Lyotard, the question he poses is also that of possible resistance against the reigning technological hegemony—the

---

72. See <http://www.interasiaschool.org/>.

product of occidental metaphysics. This is precisely the task of the postmodern, beyond its aesthetic expressions. Certain other thinkers such as Latour and Descola, who eschew the postmodern, are instead drawn to the 'non-modern' in order to address this task. However, no matter what we call it, Lyotard's question deserves to be taken up seriously once more. And as we shall see, this question converges with the inquiries of Nishitani, Mou, Stiegler, and Heidegger. If an anthropology of nature is possible and necessary in order to elaborate on non-modern modes of thinking, then the same operation is possible for technics. It is on this point that we can and must engage with contemporary European thought concerning the programme of overcoming modernity, as clearly and symptomatically exemplified, for instance, in the recent work of French philosopher Pierre Montebello, *Cosmomorphic Metaphysics: The End of the Human World.*[73]

Montebello attempts to show how the search for a post-Kantian metaphysics, hand-in-hand with the 'ontological turn' in contemporary anthropology, can lead us—Europeans, at least—out of the trap that modernity has set for us. Kant's metaphysics, as Montebello puts it, is based on limits. Kant already warned readers of the *Critique of Pure Reason* about the *Schwärmerei* or 'fanaticism' of speculative reason, and attempted to draw the boundaries of pure reason. For Kant, the term 'critique' doesn't carry a negative signification, but rather a positive one, namely that of exposing the conditions of possibility of the subject in question—the limits within which the subject can experience.

---

73.   P. Montebello, *Métaphysiques cosmomorphes, la fin du monde humain* (Dijon: Les presses du réel, 2015).

This setting of limits appears again when we consider Kant's division between phenomenon and noumenon, and his refusal to consider human beings capable of intellectual intuition, or intuition of the thing-in-itself.[74] For Kant, human beings only have sensible intuitions corresponding to phenomena. Montebello's formulation of the becoming of post-Kantian metaphysics, as exemplified in the thought of Whitehead, Deleuze, Tarde, and Latour, hinges on the attempt to overcome such a metaphysics of limits, and therefore proposes a necessary infinitisation. The political danger of the Kantian legacy is that human beings become more and more detached from the world, a process formulated by Bruno Latour as follows: 'Things-in-themselves become inaccessible while, symmetrically, the transcendental subject becomes infinitely remote from the world.'[75] Mou Zongsan's critique of Kant accords in this respect with Montebello's, though Mou proposes a different way to think about infinitisation—namely, through the reinvention of Kantian intellectual intuition in terms drawn from Chinese philosophy.

Montebello proposes that the work of Quentin Meillassoux stands out as a challenge to the limit of modernity (here a synonym for the Kantian legacy of a metaphysics of limits). One central feature of the latter that Meillassoux calls into question is what he calls 'correlationism'—the stipulation that any object of knowledge can only be thought in relation to the conditions according to which it is manifested to a subject. This paradigm, according to Meillassoux, has been predominant in

---

74.   Ibid., 21.

75.   B. Latour, *We have Never Been Modern* (Cambridge MA: Harvard University Press, 1993), 56; cited by Montebello, *Métaphysiques cosmomorphes*, 105.

Western philosophy for more than two centuries, for example in German Idealism and phenomenology. Meillassoux's question is simply this: How far can reason reach? Can reason accede to a temporality where it itself ceases to be, for example in thinking objects belonging to an ancestral era prior to the appearance of humanity?[76] Although Montebello acknowledges Meillassoux's work, at the same time he strategically portrays Meillassoux and Alain Badiou as representatives of a failed attempt to escape finitude that relies on the 'mathematical infinite'. When Montebello says 'mathematics' here, he means numerical reduction; and he jointly condemns both mathematics (in this sense) and correlationism:

> The monster with two heads simultaneously affirms a world without man, mathematical, glacial, desert, unlivable, and man without world, haunting, spectral, pure spirit. Mathematics and correlation, far from opposing each other, marry each other in funereal weddings.[77]

It is not our task here to examine Montebello's verdict against Badiou and Meillassoux. What interests us is the solution he proposes, which consists in affirming instead 'the multiplicity of relations that situate us in the world'.[78] We can understand this as a resistance against a thinking based on mathematical rationality, and which takes into consideration the history of cosmology, which we can analyse in terms of the progress of geometry in its departure from myth and its ultimate

---

76. Q. Meillassoux, *After Finitude: An Essay on the Necessity of Contingency*, tr. R. Brassier (London: Continuum, 2009).

77. Montebello, *Métaphysiques cosmomorphes*, 69.

78. Ibid., 55.

completion in astronomy. It seems to me that this type of relational thinking is emerging in Europe as a replacement for a substantialist thinking that has survived since antiquity. This is evident in the so-called 'ontological turn' in anthropology—for example in Descola's analysis of the ecology of relations—as well as in philosophy, where Whitehead and Simondon's anti-substantialist relational thinking is gaining more and more attention. Here the concept of relation dissolves the concept of substance, which becomes a unity of relations. These relations constantly weave with each other to construct the web of the world as well as our relations with other beings. Such a multiplicity of relations can be found in many non-European cultures, as demonstrated in the works of anthropologists such as Descola, Viveiros de Castro, Ingold, and others. In these multiplicities of relations, one finds new forms of participation according to different cosmologies, and in this sense Montebello proposes to think about *cosmomorphosis* rather than anthropomorphosis—to think beyond the *anthropos* and to reconfigure our practices according to the *cosmos*. Naturalism, as we have seen above, is only one such cosmology alongside others such as animism, analogism, totemism, and what Viveiros de Castro calls 'perspectivism', meaning the exchange of perspectives between human and animals (where, for example, the peccary sees itself as hunter, and vice versa). Viveiros de Castro uses Deleuze and Guattari's concept of intensity to describe a new form of participation, 'becoming-others', which sheds light on the possibilities of a post-structural anthropology. The importance of Viveiros de Castro's contribution is that he introduces a new way to do anthropology that is not confined to the legacy of Lévi-Straussian structuralism. To his eyes, if Western relativism (e.g. the recognition of multiple ontologies) implies a multiculturalism as public politics, then Amerindian perspectivism can give us a multinaturalism as cosmic politics.[79] Unlike naturalism, these other forms of cosmology operate according to continuities (e.g. intensities, becoming) rather than discontinuities between culture and nature. For the same reasons, I propose to investigate technological thinking in China without adopting the structuralist anthropological approach fashioned by sinologists such as A.C. Graham and B.I. Schwartz.

Montebello argues that a return to a more profound philosophy of nature is able to overcome the Anthropocene—the symbol of modernity—by bringing back a new way of *being together* and *being with*. Such a concept of nature is one that would resist the division between culture and nature found in naturalism. Now, the examples Montebello borrows from Descola and Viveiros de Castro resonate strongly with the concept of *Dao*, as a cosmological and moral principle which, as I discuss below, is based on the resonance between (and the unification of) the human and the Heaven. The Chinese cosmology, based on this resonance, is ultimately a moral cosmology—it is this cosmological view that defines the interaction between humans and the world, in terms of both natural resources and cultural practices (family hierarchy, social and political order, public policies, and human/non-human relations). Indeed, in the work of Descola one finds occasional references to Chinese culture, which seem to originate in the work of Jullien and Granet. Reading Granet, for example, Descola finds that during the European Renaissance, analogism rather than naturalism was the dominant ontology.[80] Naturalism,

79.   E. Viveiros de Castro, *Cannibal Metaphysics: For a Post-Structural Anthropology*, tr. P. Skafish (Minneapolis: Univocal Publishing, 2014), 60.

80.   Descola, *Beyond Nature and Culture*, 206–7.

in this sense, is only a product of modernity; it is 'fragile' and 'lacking in ancient roots'.[81]

Yet I am sceptical that this kind of return to or reinvention of the concept of 'nature', or a return to some archaic cosmology, is sufficient to overcome modernity. This scepticism is both epistemological and political. Montebello mobilises Simondon to show that nature is the 'pre-individual', and that it is therefore the foundation of all forms of individuation. It is true that Simondon speaks of

> [t]his pre-individual reality that the individual carries within it [and which] could be named nature, thereby rediscovering in the word "nature" the meaning that the pre-Socratic philosophers gave it [...] Nature is not the opposite of man, but the first phase of being, the second phase being the opposition of the individual and the milieu.[82]

But what is 'nature' for Simondon? As I have shown elsewhere,[83] the existence of two separate currents of reception of Simondon—as philosopher of nature, and as philosopher of technology, based respectively on his two theses *L'individuation à la lumière des notions de forme et d'information* and *Du mode d'existence des objets techniques*—remains problematic, since what Simondon in fact sought to do was to overcome the discontinuities between nature, culture, and technics. What is in question here is not just the interpretation of Simondon, but rather this 'nature'

---

81. Ibid., 205.

82. G. Simondon, *L'individuation à la lumière des notions de forme et d'information* (Grenoble: Jérôme Millon, 2005), 297.

83. Yuk Hui, *On the Existence of Digital Objects* (Minneapolis: University of Minnesota Press, 2016).

itself: and the tension between 'nature' and the global techno-logical condition will not disappear just because of a narrative of the 'ontological turn'.

This observation brings us to the global techno-political dimension that I would like to add to this discourse. It is understandable that a European philosopher might believe that, once Europe manages to distance itself from modernity, then other cultures will be able to resume their interrupted cosmologies; and that therefore, in opening up European thought to other ontologies, he also saves the Other from its subjection to Western technological thinking. But there is a blind spot at work here: when Montebello and others recognise that European naturalism is a rare and perhaps exceptional case, they don't seem to take account of the extent to which this view has pervaded other cultures through modern technology and colonisation. Those cultures which, over the past century, have had to contend with European colonisation, have already undergone great changes and transformations, to the extent that the global technological condition has become their own destiny. Given this 'reversal' in perspectives, any 'return to nature' is questionable at best.

This book would like to offer another standpoint, using China as *an example* to describe the 'other side' of modernity, and hopefully providing some insights into the current pro-gramme of 'overcoming modernity' or 'resetting modernity' in the era of digitalisation and the Anthropocene. To return to ancient categories and invoke the concept of cosmotechnics is by no means to return to them as 'truth' or as 'explanation'. The scientific knowledge of today confirms that many of the ancient modes of thought are replete with misconceptions, and on this basis a certain scientism may even refuse any consid-eration of the question of Being and the question of *Dao* alike.

However it should be restated that, through the trajectory that this book will outline, I seek to reinvent a *cosmotechnics*, and not just to return to belief in a cosmology. Neither do I seek a return to nature, in the sense that many read Ionian philosophy or Daoist philosophy as a philosophy of nature—but rather to reconcile technics and nature, as Simondon proposed in his thesis on the genesis of technicity.

§6. SOME NOTES ON METHOD

Before embarking upon our inquiry, a few words should be added concerning its method. Although I attempt to outline the historical transformations of the *Qi-Dao* relation, I am aware that its complexity is far beyond the simple sketch that I can offer here, since it is impossible to exhaust this dynamic in such a modest essay. The generalizations and unconventional readings that this book is obliged to carry out have to be recognized for their limits and prejudices, but there is no way to carry out such a project without working through them. Nevertheless, I hope that what I set out below will be of inspiration to scholars who might want to address the question of technology from both European and non-European perspectives—something that I believe is becoming increasingly necessary.

Rather than presenting a formal method, I would like to explain three things that I seek to *avoid*: Firstly, a symmetry of concepts, where one starts with corresponding concepts in European philosophy and Chinese philosophy—for example, identifying the equivalents of *technē* and *physis* in Chinese culture. It is true that, after decades of progress in translation and cultural communication, the terms of Western philosophy can find more or less corresponding translations in the Chinese language. But it is dangerous to take these as symmetrical relations.

For the search for symmetry will end up obliging us to use the same concepts, or more precisely, to subsume two forms of knowledge and practice under predefined concepts. To start with, asymmetry also means an affirmation of difference—but not a difference without relation (e.g. mirror images, reflections, mirages)—and to seek a convergence conditioned by this difference. Hence, in my inquiry into the question of technics in China, even if I use the word *technics*, readers should be aware of the linguistic constraints, and must be prepared to open themselves to a different cosmological and metaphysical system. For these reasons, I do not use the usual translation of *technē* as *Gong* (工, 'work') or *Ji* (技, 'skill'), which would turn our inquiries into mere empirical examples; but rather start with a systematic view of *Qi* (器) and *Dao* (道), terms which, in turn, cannot be reduced to product (*ergon*) and soul (*psychē*). This asymmetry is presupposed and methodologically mobilised in this book. Readers may find that on occasion I try to draw out similarities, but only so as to render visible the underlying asymmetry.

The same thing goes for translating doctrines such as dualism and materialism. For example, it would be incorrect to understand *Yin-Yang* as a dualism in the same sense that we use this term in Europe. The latter generally refers to two opposing and discontinuous entities: mind-body, culture-nature, being-nothingness. This form of dualism is not dominant in China, and *Yin-Yang* are not conceived of as two discontinuous entities. Hence in Chinese metaphysics there is virtually no problem in recognizing that being comes from nothingness, as is already stated in the Daoist classics. In Europe, ex nihilo creation is the reserve of a divine power, since it is scientifically impossible: *ex nihilo nihil fit*. It was not until Leibniz posed the question 'Why is there something instead of nothing?', later taken up by Heidegger to explicate the meaning of Being, that the question of Being would be further clarified in Western philosophy. In more general terms, Chinese thinking tends to be concerned more with continuity and less with discontinuity. This continuity is constructed by relations, as found, for example, in resonances between the Heaven and the human, musical instruments, or the moon and flowers. As mentioned above, this is often referred to as 'correlative thinking'.[84] However, this discourse is developed by Granet and later by A.C. Graham, who make use of structuralist anthropology to formulate the two corresponding entities as oppositions, for example *Yin-Yang*. I prefer to call it a 'relational' rather than 'correlative' thinking, because the correlative thinking described by the above mentioned sinologists inspired by structuralist anthropology is always mobilised in an attempt to systematise, in order finally to present static structures.[85] This relational thinking is in fact more open than this might suggest, since it is more dynamic. It does indeed include a correlative mode of association, meaning that one natural phenomenon can be related to the other according to shared common categories in the cosmology—for example, the *Wu Xing* ('five phrases', or 'five movements'). But it can also be political, in the sense that there is a correlation between seasonal change (as the expression of the will of the Heaven) and the policy of the state—for example, one should avoid executing criminals in the springtime. Finally, it can also be subtle and poetic, in the sense

---

84. See Graham, *Yin Yang and the Nature of Correlation*, chapter 2.

85. Readers interested in how a structuralist reading is performed can refer to B.I. Schwartz, *The World of Thought in Ancient China* (Cambridge, MA: Harvard University Press, 1985), chapter 9: 'Correlative Cosmology: The "school of *Yin* and *Yang*"', where Schwartz analyses the school using a method similar to Lévi-Strauss's primitive 'science of the concrete'.

that the heart is able to detect this subtle resonance between natural phenomena in order to reach the *Dao*—something that is especially true in the Xin school of Neo-Confucianism.

Secondly, I avoid setting out from isolated concepts as if they were static categories—a method practiced by many sinologists, but which seems to me rather problematic, because it also unconsciously imposes a sort of cultural essentialism. Concepts can never exist independently: a concept exists in relation to other concepts; moreover, concepts are transformed over time, either in themselves or in relation to a broader system of concepts. This is especially so in Chinese thinking—which, as we have said, is fundamentally a relational thinking. Therefore, instead of comparing two concepts, I try to take a systematic view and to open up the possibility of locating a *genealogy* of the concept within the system. As we shall see, when we focus on the relation between *Dao* and *Qi*, we must consider both their historical separation and their reunification as the lineage through which we can project a philosophy of technology in China. I hope that the case of China can serve as an example to illustrate this difference, and hence contribute toward a pluralism of technicity.

Thirdly, I would like to distance this work from postcolonial critiques. This is not at all to say that postcolonial theory is not taken into account here, but rather that I aim to provide a supplement that makes up for what postcolonial theory tends to disregard. The strength of postcolonial theory, it seems to me, is that it effectively reformulates the question of power dynamics as narratives, and consequently argues for other, or different, narratives. However, this might also be regarded as one of its weaknesses, since it tends to ignore the question of technology—a question which, I would argue, cannot be reduced to one of narratives. Indeed, it is dangerous to try to operate such a reduction, since doing so involves acknowledging the material conditions without understanding the material significance of these conditions—just as *Qi* was considered to be inessential to *Dao* during the social and political reform in China after the Qing dynasty (see §14). Thus the approach adopted here departs from that of postcolonial critique in order to advance towards a materialist critique. This materialism is not one that opposes spirit and matter, though; rather, it aims to foreground material practice and material construction in order to attain a cosmological and historical understanding of the relation between the traditional and the modern, the local and the global, the Orient and the Occident.

PART 1:
IN SEARCH OF
TECHNOLOGICAL THOUGHT
IN CHINA

§7. DAO AND COSMOS: THE PRINCIPLE OF THE MORAL

The Chinese had already dedicated a classical text to the question of technics during the Ionian period (770–211 BC). In this text, not only do we find details of various technics—wheel-making, house-building and the like—but also the first theoretical discourse on technics. In the classic in question, the *Kao Gong Ji* (考工記, *A Study of Techniques*, 770–476 BC), we read:

> Provided with the timing determined by the heavens, energy [氣, *ch'i*] provided by the earth, and materials of good quality, as well as skilful technique, something good can be brought forth through the synthesis of the four. [天有時，地有氣，材有美，工有巧。合此四者，然後可以為良]

According to this text, then, there are four elements which together determine production. The first three are given by nature, and hence are not controllable. The fourth, technique, is controllable—but it is also *conditioned* by the other three elements: timing, energy, and material. The human is the last element, and its way of being is situational. Moreover, technique is not given; it is something that has to be learned and improved.

The Aristotelians, of course, also have their four causes: formal cause, material cause, efficient cause, and final cause. For them, production starts with the form (*morphē*) and ends with the realisation of this form in the matter (*hylē*). But Chinese thought has in effect already jumped over the question of form to arrive at the question of 'energy' (*ch'i*, which literally means 'gas'); and technics is not the determining factor, but rather serves as a facilitation of *ch'i*. In this energetic view of the world, beings are joined together in a cosmic order which communicates through a consciousness common to them all; and technics concerns the ability to 'skilfully' bring something together that resonates with this cosmic order—which, as we shall see, is ultimately a *moral* order.

In 'The Question Concerning Technology', Heidegger reiterated Aristotle's four causes, and related efficient cause to the possibility of revelation. In the Heideggerian conception of the four causes, technics is in itself *poeisis* (as both production and poesy). This concept of technics may seem similar to the Chinese one, but there is a fundamental difference: unlike the Chinese concept of technics as realising the 'moral good' of the cosmos, Heidegger's interpretation of Aristotle's technics reveals 'truth' (*alētheia*), the unconcealment of Being. Of course, what Heidegger understands as truth is not logical truth, but rather a revelation of the relation between *Dasein* and its world, a relation usually ignored in the perception of the world as present-at-hand. Nonetheless, the pursuit of the moral and the pursuit of truth characterise divergent tendencies of Chinese and Greco-German philosophy. Both Greece and China had their cosmologies, which would in turn leave their mark on their respective cosmotechnical dispositions. As the philosopher Mou Zongsan (1909–1995) insisted, the Chinese cosmology was a moral ontology and a moral cosmology, meaning that it did not originate as a philosophy of nature, but as a moral metaphysics, as is stated in the Qian (乾文言) of the *I Ching*:

> The moral of the great man is identical with that of Heaven and
> Earth; his brilliance is identical with that of the sun and the moon;

his order is identical with that of the four seasons, and his good and evil fortunes are identical with those of spiritual beings.[1]

What is meant by 'moral' in the Confucian cosmology has nothing to do with heteronomous moral laws, but concerns creation (which is exactly the meaning of *Qian*) and the perfection of personality. For this reason, Mou distinguishes Chinese moral metaphysics from a metaphysics of morals, for the latter is only a metaphysical exposition of the moral, while for Mou, metaphysics is only possible on the basis of the moral.

Compared to Presocratic and classical Greek philosophy, during the same period of history in China, neither the question of Being nor the question of *technē* comprised the core questions of philosophy. Rather than 'Being', what was common to Confucian and Daoist teaching was the question of 'Living', in the sense of leading a moral or good life. As François Jullien attempted to show in his *Philosophie du vivre*, this tendency led to a totally different philosophical mentality in China.[2]

---

1. '大人者與天地合其德，與日月合其明，與四時合其序，與鬼神合其凶'. *I Ching*, Qian Gwa (乾・文言); quoted by Mou Zongsan, *The Questions and Development of Sung and Ming Confucianism* (宋明儒學的問題與發展) (Shanghai: Huadong Normal University Press, 2004), 13.

2. F. Jullien, *Philosophie du Vivre* (Paris: Gallimard, 2011). This having been said, I am conscious that it requires some justification, since it will inevitably elicit disagreements concerning the interpretation of the history of Western philosophy. It is true that in this book I address Heidegger's reading of the history of metaphysics; but I do not wish to ignore the fact that in the Hellenistic schools (e.g. Cynics, Epicureans, and Stoics) and their Roman continuations there was a whole tradition of *technē tou biou*, or 'technologies of the self', to use Foucault's term (M. Foucault, 'Technologies of the Self', in L.H. Martin, H. Gutman and P.H. Hutton [eds], *Technologies of the Self: A Seminar with Michel Foucault*, [Amherst, MA: University of Massachusetts Press, 1988], 16–49). (The Hellenistic schools' emphasis on the care of the self seems to resonate with what Heidegger calls *Sorge* in *Being and Time*; indeed, he cites Seneca's *Epistulae morales ad Lucilium* when he talks about *cura* in §42. Victor Goldschmidt has argued that Heidegger's distinction between physical time

It is undeniable that there were certain philosophies of nature in ancient China, notably in Daoism and its further 'technical' continuation in alchemy. But this philosophy of nature did not indulge in speculations on the basic material elements of the world, as was the case with Thales, Anaximander, Empedocles, and others, but rather treated of an organic or synthetic form of life—organic in the sense of its being subject to a mutual causation in which the universe is considered as a totality of relations.[3] In Confucianism, *Dao* is recognised as the coherence between the cosmological and moral orders; this coherence is called *zi ran* (自然), which is often translated as 'nature'. In modern Chinese the term refers to the environment, to the wild animals, plants, rivers, etc. that are already given; but it also means acting and behaving according to the self without pretention, or letting things be as they are. This self, however, is not a *tabula rasa*, but emerges out of, and is nourished and constrained by, a certain cosmic order, namely *Dao*. In Daoism, on the other hand, '*Zi ran* is the law of *Dao* (道法自然)' was

---

and living time cannot be applied to the Stoics, since they have a different concept of *physis*, something we address below (§10.3). See V. Goldschmidt, *Le système stoïcien et l'idée du temps* [Paris: Vrin, 1998], 54.) It is always puzzling to see how beautifully Heidegger circumscribed Hellenistic philosophy, yet saw Roman philosophy as consisting merely in poor translations of ancient Greek philosophy. Is it because the question of Being was not evident among the Hellenistic thinkers, or is this episode simply incompatible with his history of Being? One might even speculate that there is a certain incompatibility between the Stoic cosmotechnics and Heidegger's definition of *technē*. These are questions that deserve to be addressed in rigorous detail. For present purposes, I will restrict myself to the discussion of the metaphysical inquiry, against the backdrop of Heidegger's essay on technology, but I will come back to the Stoics and Daoists in §10.3.

3.　In the Chinese mythology of creation, the universe was formed when a giant called Pang Gu divided the primal chaos into heaven and earth with an axe; after his death, his body parts were transformed into the mountains (bones) and rivers (intestines).

both the slogan and the principle of a philosophy of nature.[4]
These two concepts of *Dao* in Confucianism and Daoism have
an interesting relation to one another, since on the one hand,
according to conventional readings, they seem to be in tension:
Daoism (in the texts of Laozi [−531 BC] and Zhuangzi [370−287
BC] is very critical of any imposed order, whereas Confucianism
seeks to affirm different kinds of order; on the other hand, they
seem to supplement each other, as if one asks after the 'what',
the other after the 'how'. As I will argue below, however, they
both embody what I call a 'moral cosmotechnics': a relational
thinking of the cosmos and human being, where the relation
between the two is mediated by technical beings. It is therefore
not my intention to read these relations between *Dao* and
beings as a philosophy of nature, but rather to understand
them as a possible philosophy of technology in both Confucianism and Daoism. According to this parallel reading, then,
in Chinese philosophy *Dao* stands for the supreme order of
beings; and technique must be compatible with *Dao* in order to
attain its highest standard. Accordingly, this highest standard
is expressed as *the unification of Dao and Qi* (道器合一). As
we noted in the Introduction, in its modern sense *Qi* means
'tool', 'utensil', or more generally, 'technical object'. Early Daoists
such as Laozi and Zhuangzi believed that the 'ten thousand
beings' (*wan wu*, 萬物) emerge through *Dao*; as Laozi writes,

---

4.   Laozi, *Tao de ching*. The translation of Stephen Addiss and Stanley
Lombardo is 'Dao follows its own nature'. It is possible to understand the
phrase like this; it is debatable, however, since 'nature' here suggests 'essence',
whereas *Dao* has no essence—see Lao-Tzu, *Tao Te Ching* (Indianapolis:
Hackett, 1993), 25.

Dao engenders One, One engenders Two, Two engenders Three, Three engenders the Ten Thousand beings. [道生一，一生二，二生三，三生萬物][5]

Hence *Dao* is present in thousands of beings as *de* (德, 'virtue'), and in such forms is not separated from beings; it is immanent. The usual translation of *de* as 'virtue' is debatable, however, since in *Tao te ching* (or *Laozi*), *de* doesn't have the connotation of virtue or moral perfection, but rather signifies the original harmony of the productive force of the cosmos.[6] *Dao* is present everywhere and in every being, as Zhuangzi said, since that which creates beings is not separate from those beings (物物者與物無際). For Zhuangzi, the presence of *Dao* in being takes the form of *ch'i* (氣: as mentioned above, the word literally means 'gas', but is often translated as 'energy').[7] This relation between *Dao* and Being, or *Dao* and *ch'i* was made explicit by the scholar of the Wei Jin dynasties Wang Bi (王弼, 226–249), whose commentary on *Laozi* served as the basis for the study of Daoism for centuries, prior to the discovery of the earliest version.[8] Wang Bi derived four analogical pairs, each of which is considered to stand in

---

5. Ibid., 42.

6. E.T. Ch'ien (錢新祖), *Lectures on the History of Chinese Thought* (中國思想史講義) (Shanghai: Orient Publising Center, 2016), 127. Ch'ien argues that in paragraph 55 of *Dao de Ching*, when Laozi says 'one who preserves *de* in fullness is to be compared to an innocent infant (含德之厚，比於赤子)', *de* refers not to virtue but rather to *zi ran*.

7. Chen Guu Ying (陳鼓應), 'On the Relation between Tao and Creatures: The Main Thread in Chinese Philosophy [《論道與物關係問題：中國哲學史上的一條主線》]', 台大文史哲學報 62 (May 2005), 89–118: 110–12.

8. Two earlier versions were discovered at the archaeological sites of Mawangdui (1973) and Guodian (1993). The Guodian bamboo slips are considered to be the earliest extant version, and contain several differences to Wang Bi's version.

a similar relation: (1) *Dao-Qi* (道器); (2) Nothing-Being (無有); (3) Centre-Periphery (本末); and (4) Body-Instrument (體用).[9] The unity of each pair embodies the holistic view of Chinese philosophy. There has been consensus on this point: even though there is a difference, for example, between *Dao* and *Qi*, they cannot be separated as if they were two entities.

In the section 'Knowledge Wandered North', Zhuangzi, like Spinoza, announces that *Dao* is omnipresent:

> Master Dongguo asked Zhuangzi, 'This thing called the *Dao*—where does it exist?'
> Zhuangzi, said, 'There's no place it doesn't exist.'
> 'Come,' said Master Dongguo, 'you must be more specific!'
> 'It is in the ant.'
> 'As low a thing as that?'
> 'It is in the panic grass.'
> 'But that's lower still!'
> 'It is in the tiles and shards.'
> 'How can it be so low?'
> 'It is in the piss and shit!'
> Master Dongguo made no reply.[10]

One might easily conclude from this that this conception of *Dao* entails a philosophy of nature. Furthermore, although it may seem a surprising anachronism, this philosophy of nature would have less affinity with what we know of Ionian philosophy than with what appeared much later in Kant, Schelling,

9.   Chen Guu Ying, 'On the Relation between Tao and Creatures', 113.

10.   Zhuangzi, *The Complete Works of Zhuangzi*, tr. B. Watson (New York: Columbia University Press, 2013), 182 [translation modified].

and other early Romantics—namely, their thinking of organic form. In §64 of his *Critique of the Power of Judgement*, Kant instigated a new inquiry into the question of organic form, which differs from the mechanical submission to a priori categories; unlike the latter, it insists on the part-whole relation of the being, and the reciprocal relations between part and whole. Kant was alerted to this question through his researches into the natural sciences of his time, and the concept of organic form would be further developed by the early Romantics. But this conception of life, nature, and cosmos as an organic being resonates strongly with the modern interpretation of Daoist thinking, for which it functions as the principle of every being.

Furthermore, *Dao* is not a particular object, nor is it the principle of a specific genre of objects; it is present in every being, yet escapes all objectification. *Dao* is *das Unbedingte*, the 'unconditioned' common to the Idealist projects of the nineteenth century that sought to find the absolute foundation of the system, that is to say a first principle (*Grundsatz*) that is wholly self-dependent. For Fichte this was the I, which is the possibility of such an unconditioned; in Schelling's early *Naturphilosophie*, it moved from the I (when he was still a follower of Fichte, from 1794 to 1797) to Nature (1799, in the *First Outline of a System of the Philosophy of Nature*). In the *First Outline*, Schelling takes up Spinoza's distinction between *natura naturans* and *natura naturata*, understanding the former as the infinite productive force of nature, and the latter as its product. *Natura naturata* emerges when the productive force is hindered by an obstacle, just as a whirlpool is produced when the current encounters an obstacle.[11]

---

11.   F.W.J. von Schelling. *First Outline of a System of the Philosophy of Nature*, tr. K.R. Peterson (New York: State University of New York Press, 2004), 18.

Thus the infinite is inscribed in the finite being, like the world soul described by Plato in the *Timaeus*, which is characterised by a circular movement.[12] We see a further continuation of the philosophy of organism in Whitehead's writings, which found great resonance in early twentieth-century China.[13] Understood in this way, *Dao* is the unconditioned that founds the conditioned perfection of all beings, including technical objects. Certainly, as Dongguo Zi imagined, *Dao* must exist in the most superior forms or objects in the world; however, as we have seen, Zhuangzi shattered his lofty illusions by placing *Dao* also in the most inferior and even undesirable objects of human life: ants, panic grass, earthenware tiles, and finally excrement. The pursuit of *Dao* resonates with what Confucius calls 'the Principle of the Heaven' (天理), a phrase also used by Zhuangzi. In this specific instance, the natural and the moral meet, and the two teachings converge on this point: to live is to maintain a subtle and complicit relation with *Dao*, even without fully knowing it.

§8. TECHNE AS VIOLENCE

As this surprising analogy with German Idealism demonstrates, although there are certain similarities across the two cultures, the concepts of nature and technics, as well as the relation

12.   If I refer here to Schelling rather than Plato, one of the reasons is that in the early Schelling's concept of nature, and as in Daoism, there is no role for the Demiurge.

13.   The question of which Western model is closer to the Chinese one remains debatable. For example, Mou Zongsan and Joseph Needham both refer to Whitehead when they talk about the essence of Chinese thinking; however, I believe that further research is needed here, and the relation between Whitehead and Schelling (which appears, for instance, in Whitehead's *The Concept of Nature* [Cambridge: Cambridge University Press, 1920], 47, where Whitehead invokes Schelling to support his argument) remains to be elucidated.

between the two, are significantly different in early Greek and Chinese thinking. The Greek word *physis* refers to 'growth', to 'bringing forth',[14] 'the process of natural development,'[15] the roman translation also carrying the connotation of 'birth',[16] while *zi ran* doesn't necessarily carry this connotation of productivity—it also applies to decay or stasis. For the ancient Greeks, technics imitates nature and at the same time perfects it.[17] *Technē* mediates between *physis* and *tychē* (chance or coincidence). This idea that technics might supplement and 'perfect' nature could not possibly occur in Chinese thought, since for the latter technics is always subordinate to the cosmological order: to be part of nature is to be morally good, since it implies a cosmological order which is also a moral order. Moreover, for the Chinese, there is certainly chance, but it is not the opposite of technics, nor is it to be overcome through technics: for chance is a part of *zi ran*, and hence one can neither resist nor overcome it. Nor is there need of any violence in order to reveal the truth, as claimed by Heidegger of the ancient Greek conception; one can only embody the truth through harmony, rather than discovering it through external means, as is the case for *technē*.[18]

---

14. W. Schadewaldt, 'The Greek Concepts of "Nature" and "Technique"', in R.C. Scharff and V. Dusek (eds), *Philosophy of Technology: The Technological Condition, An Anthology* (Oxford: Blackwell, second edition 2014), 26.

15. C.H. Kahn, *Anaximander and the Origins of Greek Cosmology* (New York: Columbia University Press, 1960), 201. Kahn further points out that 'nature' and 'origin' are united in one and the same idea.

16. P. Aubenque, 'Physis', *Encyclopædia Universalis*, <http://www.universalis.fr/encyclopedie/physis/>.

17. Schadewaldt, 'The Greek Concepts of "Nature" and "Technique"', 30.

18. This difference may also explain why there was no equivalent to the Greek concept of tragedy in ancient China: *tychē*, according to scholars such as Martha Nussbaum, is the fundamental element of Greek tragedy.

Heidegger characterises such necessary violence as the metaphysical meaning of *technē*, and of the Greek conception of the human as technical being. As early as 1935, in the lecture *Introduction to Metaphysics*, Heidegger develops an interpretation of Sophocles's *Antigone* which is also an attempt to resolve the opposition between the philosophy of Parmenides and Heraclitus, a thinker of being versus a thinker of becoming.[19]

What is striking in Heidegger's reading in *Introduction to Metaphysics*, as made explicit by Rudolf Boehm,[20] is that

---

Unavoidable chance interrupts the order of nature, and hence chance becomes the necessity of tragedy—for example, in the case of the ingenious Oedipus who, even though he solves the riddle of the Sphinx, is not able to avoid his foretold destiny; indeed, his victory over the Sphinx only paves the way to this destiny, leading him to become king and to marry his mother. See M. Nussbaum, *The Fragility of Goodness: Luck and Ethics in Greek Tragedy and Philosophy* (Cambridge: Cambridge University Press, 2001).

19.   Recall that in Heidegger's reading of Parmenides and Heraclitus, the fundamental question concerns the interpretation of the word *logos*, which comes from the verb *legein*, and for Heidegger essentially means 'letting-lie-before', 'bring-forward-into-view', the presencing of the presence as truth, *alētheia*. Parmenides's *moira* (fragment 8: 'since Moira bound it (being) to be whole and immovable'), the deity of earth, is *physis*, whose constantly coming into presence is the *logos*, see M. Heidegger, 'Moira Parmenides VIII, 34-41', in *Early Greek Thinking*, tr. D. F. Krell and F. A. Capuzzi (San Francisco: Harper, 1985), 97. In the interpretation of the *alētheia* of Heraclitus in 'Alētheia (Heraclitus Fragment B 16)' in *Early Greek Thinking*, Being is in constant self-revealing and self-concealing or hiding, as it is indicated in Fragment 123 that 'the essence of things likes to hide', which Heidegger translates into 'rising (out of self-concealing) bestows favor upon self-concealing' (114). Heraclitus's fire is the 'lighting' (*Lichtung*) that illuminates what is present, and brings them together to prepare for the presencing. Mortals may remain forgetful towards the lighting, since they are concerned only with what is present (122). The appropriation of the revealing-concealing of Being as event (*Ereignis*) is presented as the *logos*.

20.   R. Boehm, 'Pensée et technique. Notes préliminaires pour une question touchant la problématique heideggerienne', *Revue Internationale de Philosophie* 14:52 (2) (1960), 194–220: 195.

techné constitutes the origin of thinking. This is at odds with the conventional interpretation of Heidegger's work, according to which the question of Being is an exit from the history of metaphysics, identified with the history of technics that began with Plato and Aristotle. In the *Introduction to Metaphysics*, Heidegger points out that in the first cited strophe, the human is *to deinotaton*, the uncanniest of the uncanny (*das Unheimlichste des Unheimlichen*): 'Manifold is the uncanny, yet nothing uncannier than man bestirs itself, rising up beyond him' (το δεινον is translated by Hölderlin as *Ungeheuer* ['monstrous']; Heidegger's reading conflates the three words into one: *Unheimlich*, *Unheimisch* ['homeless'] and *Ungeheuer*.)[21] For the ancient Greeks, according to Heidegger, *deinon* traverses the opposed con-frontations of Being (*Auseinander-setzungen des Seins*). The tension between being and becoming is the fundamental element here. According to Heidegger, the uncanny is said in two senses: it is firstly said of the violence (*Gewalttätigkeit*), the act of violence (*Gewalt-tätigkeit*), in which the essence of human being as *techné* consists: human beings are Daseins that overstep the limits; in so doing, the Dasein of the human being finds itself no longer at home, it becomes *unheimlich*.[22] This violence associated with *techné* is neither art nor technics in the modern sense, but knowing—a form of knowing that can set Being to work in beings.[23] Secondly, it is said of overwhelming (*Überwaltigend*) powers such as that of the sea and the earth.

---

21.   Heidegger, *GA 53. Hölderlins Hymne 'der Ister'* (Frankfurt am Main: Vittorio Klostermann, 1993), 86.

22.   Heidegger, *GA 40. Einführung in die Metaphysik* (Frankfurt am Main: Klostermann, 1983), 116.

23.   Ibid., 122.

This overwhelming is manifested in the word *dikē*,[24] which is conventionally translated as 'justice' (*Gerechtigkeit*). Heidegger translates it as 'fittingness' (*Fug*), since *iustitia*, the Latin word for justice, 'has a wholly different ground of essence than that of *dikē*, which arises from [*west*] *aletheia*':[25]

> We translate this word [*dikē*] as fittingness [*Fug*]. Here we understand fittingness first in the sense of joint [*Fuge*] and structure [*Gefüge*]; then as arrangement [*Fügung*], as the direction that the overwhelming [*Überwältigende*] gives to its sway; finally, as the enjoining structure [*fügende Gefüge*], which compels fitting-in [*Einfügung*] and compliance [*sich fügen*].[26]

The play on the word *Fuge* and its derivations—*Gefüge*, *Fügung*, *fügende Gefüge*, *Verfügung*, *Einfügung*, *sich fügen*— is totally lost in the English translation. These cognates make it clear that *dikē*, normally translated as 'justice' in the legal and moral sense, is for Heidegger firstly a joint, a structure; and then an arrangement that is directed toward something—but who is directing it? *Glückliche Fügung* is often translated as 'fortunate coincidence', yet it is not a totally contingent happening, but rather one that is born of external forces. And finally, it is a compelling force, to which the compelled has to

24. Hesiod tells us in his *Theogony* that Zeus married Themis and sired the daughters Horae (Hours), Eunomia (Order), Dikē (Justice) and Eirene (Peace); and according to Orpheus, 'Dikē sits next to Zeus's throne and arranges all human affairs', F. Zore, 'Platonic Understanding of Justice: On Dikē and Dikaiosyne in Greek Philosophy', in D. Barbarić (ed.), *Plato on Goodness and Justice* (Cologne: Verlag Königshausen & Neumann, 2005), 22.

25. C.R. Bambach, *Thinking the Poetic Measure of Justice: Holderlin-Heidegger-Celan* (New York: SUNY Press, 2013), 34; Heidegger, *GA 54*, 59. *Hölderlins Hymne 'Andenken'* (winter semester 1941/42) (Frankfurt am Main: Vittorio Klostermann, 1982). 59.

26. Heidegger *GA 40*, 123; *Introduction to Metaphysics*, tr. G. Fried and R. Polt (New Haven, CT: Yale University Press, 2000), 171.

submit in order to be part of the structure. It is at this moment that we can observe the opposition between *technē* and *dikē*, the 'violent act' of the Greek *Dasein* and the 'excessive violence of Being [*Übergewalt des Seins*]'.[27] 'Violent acts' such as language, house building, sailing, etc., Heidegger emphasises, shouldn't be understood anthropologically, but rather in terms of mythology:

> The violence-doing of poetic saying, of thoughtful projection, of constructive building, of state-creating action, is not an application of faculties that the human being has, but is a disciplining and disposing of the violent forces by virtue of which beings disclose themselves as such, insofar as the human being enters into them.[28]

This confrontation is for Heidegger the attempt to open up withdrawn Being according to the Presocratics. It is a necessary confrontation, since 'historical humanity's Being-here means: Being-posited as the breach into which the excessive violence of Being breaks in its appearing, so that this breach itself shatters against Being.'[29] In this *theatre of violence*, the human's assault on Being comes out of an urgency necessitated by Being, by the holding sway of *physis*. This *Auseinandersetzung* between *technē* and *dikē* can be understood, according to Heidegger, as the 'being as a whole' of Parmenides, to which both 'thinking' and 'being' belong; but it also perfectly accords with the teaching of Heraclitus, according to which 'it is necessary to keep in view confrontation,

---

27. Ibid., 124; *Introduction to Metaphysics*, 173.

28. Ibid., 120; *Introduction to Metaphysics*, 167.

29. Bambach, *Thinking the Poetic Measure of Justice*, 174.

setting-apart-from-each-other [*Aus-einander-setzung*] essentially unfolding as bringing-together, and fittingness as the opposed...'.[30] This confrontation is a disclosure of Being as *physis*, *logos* and *dikē*, and sets Being to work in beings; consequently, Heidegger concludes, 'The overwhelming, Being confirms itself in works *as history*.'[31]

Neither *dikē* nor *nōmos*, as Vernant points out, had an absolute systematic connotation for the ancient Greeks. For example, in *Antigone*, what Antigone calls *nōmos* is not the same as what Creon understands by the term.[32] The translation of *dikē* as *Fug* (fittingness) invoked in the *Introduction to Metaphysics* is taken up again in 1946 in the essay *Der Spruch des Anaximander*. Here Heidegger argues against the translation, suggested by Nietzsche and the classical scholar Hermann Diels, of *dikē* as *Buße* or *Strafe* ('penalty'), and once again proposes instead to translate *dikē* as *Fug*, the ordering and enjoining order (*fugend-fügende Fug*),[33] and *Adikia* as *Un-Fug*, disjunction, disorder. Nietzsche's translation is as follows:

---

30.   Heidegger, *Introduction to Metaphysics*, 177, citing Heraclitus, Fragment 80. The sentence is conventionally translated as 'But it is necessary to know that war is common to all and justice is strife.' This opposition can probably be more easily understood when we refer to two passages in Heraclitus's fragments: in fragment B51, 'they do not understand in what manner what is differentiated [*diapheromenon*] concurs with itself: a framework [*harmoniē*] consisting in opposing tensions [*palintropos/palintonos*], such as that of a bow or of a lyre'; and in B53, where it is expressed in an even more violent way: 'War is the father of all things and king of all things: some it appointed as gods, some as human beings, some it made into slaves and some into free men', cited by J. Backman, *Complicated Presence: Heidegger and the Postmetaphysical Unity of Being* (New York: SUNY Press, 2015), 32, 33.

31.   Ibid., 125; *Introduction to Metaphysics*, 174.

32.   J.-P. Vernant and P. Vidal-Naquet, *Myth and Tragedy in Ancient Greece*, tr. J. Lloyd (New York: Zone, 1990), 26.

33.   M. Heidegger, GA 5. *Holzwege (1935–1946)* (Frankfurt am Main: Klostermann, 1977), 297; Heidegger, *Early Greek Thinking*, 43.

'Whence things have their origin, there they must also pass away according to necessity; for they must pay penalty and be judged for their injustice, according to the ordinance of time.'[34] Heidegger's reinterpretation of Anaximander's fragment is an attempt to retrieve the history of Being, which is arriving at an abyss. As readers of Heidegger will know, the ontological difference between Being (*Sein*) and beings (*Seiendes*) and their dynamics constitute a history of occidental metaphysics in which the forgetting of Being and the presence of beings as totality lead to what he calls the 'eschatology of Being'.[35] Beings as mere presence are in disorder, are out of joint; hence Heidegger renders Nietzsche's translation of the second part of the fragment as 'they let order belong (*didōnai…dikēn*), and thereby also reck, to one another (in the surmounting) of disorder'.[36] Heidegger deliberately links the word reck (*Ruch*, whose original sense can no longer be recovered) with order, *dikē*. He also mentions the Middle High German word *ruoche*, which means solicitude (*Sorgfalt*) and care (*Sorge*), without further commentary.[37] The disorder is surmounted in order to bring orders into being—the presencing of presence. It is an attempt to recover the experience of Being as the revelation of such overwhelming fittingness, rather than determining them into beings as mere presence. The point we wish to emphasise here is the necessity of revealing the *dikē* of Being through the

---

34. 'Woraus aber die Dinge das Entstehen haben, dahin geht auch ihr Vergehen nach der Notwendigkeit; denn sie zahlen einander Strafe und Buße für ihre Ruchlosigkeit nach der fest-gesetzten Zeit.' Heidegger, *Early Greek Thinking*, 13; GA 5, 297.

35. Ibid., 18. Here 'eschatology of Being' does not have a theological meaning. Instead, Heidegger claims, one should take it in the sense of a 'phenomenology of spirit'.

36. Ibid., 47.

37. Heidegger, *GA 5*, 360.

violence of *technē*. The Heidegger of 1946 no longer spoke of the violence of technics as he did in 1936, but used a much milder word, *verwinden* ('surmount'), and turned to 'poetising' on the riddle of Being. However, this poetising was not an abandonment of technics, but rather consisted in returning to technics as *poeisis*.

What Heidegger's analysis begins to suggest, then, is that the Greek relation to technics emerges from a cosmology, and that knowledge of technics is a response to the cosmos, an attempt to 'fit' or to strive for 'fittingness', or perhaps 'harmony'.[38] What characterizes this fittingness? In particular, a parallel reading of Heidegger's reading of Anaximander as a philosopher of Being and Vernant's interpretation of Anaximander as a social-political thinker reveals something peculiar regarding the role played here by the Greek 'cosmotechnical' relation to geometry. For if we refer to the ancient Greek moral theory, law (*nōmos*) is closely related to *dikē* in a geometrical sense. *Dikē* means something can be fitted into the divine order, which suggests a geometrical projection:

The *nomoi*, the body of rules introduced by the legislators, are

---

38.   This cosmological perspective is discussed but not thematised in Heidegger's seminar on Heraclitus (1966–1967), M. Heidegger, GA 15 *Heraklit Seminar Wintersemester 1966/1967* (Frankfurt am Main: Klostermann, 1986). In the 7th seminar, Heidegger posed the question concerning the difference between fragment 16 and fragment 64. Fragment 64 starts with the lightning (*Blitz*) and throughout the discussion on the all (*tā pānta*), there is no mention of the human. The relation between lightning, sun, fire, war and *tā pānta* implies a belonging-together. However, as Heidegger mentions, the difficulty is that there is a multiplicity or a manifold which *exceeds* the totality of *tā pānta* (*Andererseits ist von einer Mannigfaltigkeit die Rede, die über die Totalität hinausgeht*) (125). The puzzle is that the 'all' in which beings are conceived as totality is a *metaphysical* concept. Heraclitus's thinking is not-yet-metaphysics and no-longer-metaphysics, while *tā pānta* as a metaphysical concept marks the break between Socrates and the Ionian philosophers, as coincidentally announced by Hegel (129).

presented as human solutions aimed at obtaining specific results: social harmony and equality between citizens. However, these *nomoi* are only considered valid if they confirm to a model of equilibrium and geometric harmony of more than human significance, which represents an aspect of divine *dikē*.[39]

What Vernant reveals here is a correlation between cosmology and social philosophy in Anaximander's thought. For Anaximander, the earth is immobile (in contrast to the cosmology outlined by Hesiod in *Theogony*, in which the earth is floating) because it is at the middle (*meson*) and is balanced by other forces. The concept of the *apeiron*, the unlimited, according to Anaximander, is not an element, as it was the case with water for Thales; otherwise it would overcome or destroy all other elements.[40] Vernant gives us his interpretation of *to kratos* here: although *kratein* principally conveys a sense of domination, in Anaximander's cosmology it also denotes a supporting and a balancing. Being as whole, as one, is the most powerful; and the only possible way to ensure egalitarian relations between different beings is to impose *dikē*:

> So the rule of the *apeiron* is not comparable with a *monarchia* like that exerted by Zeus according to Hesiod, or by air and water according to the philosophers who give these elements the power to *kratein* the whole universe. The *apeiron* is sovereign in the manner of a common law that imposes the same *dikē* on each individual, that keeps each power within the limits of its own domain […].[41]

---

39. Vernant, *Myth and Society*, 95.

40. Vernant, *Myth and Thought*, 229.

41. Ibid., 231.

This relation is expressed, for example, in ancient Greek urban development, where the *agora* is placed as the heart of the city, with a circular contour—bearing in mind that the circle is the most perfect geometrical form. The *agora*, like the earth that lies in the middle (*meson*), brings about a geometrical imaginary of power: a power that does not belong to any single being, such as Zeus, but to all. Hippodamos, an architect who lived a century after Anaximander, reconstructed the destroyed Miletos according to a plan that aimed to rationalise urban space, like a checkerboard 'centred around the open space of the agora'.[42]

This synthesis of Heidegger's understanding of the original meaning of technics in relation to the *dikē* of Being and Vernant's analysis of the relation between social structure and geometry, points to the fact that geometry was foundational for both technics and justice—and we shouldn't forget that geometry was considered essential training in the school of Thales. Kahn reminds us that for both Anaximander and Pythagoras, 'the ideas of geometry are embedded in a much larger view of man and of the cosmos'.[43] This fittingness is not given as such; it is revealed only in the confrontation between the overwhelming of Being and the violence of *technē*. So should we see Heidegger's return to the original *technē* as a quest for the spirit of the ancient Greek cosmotechnics?[44]

---

42. Ibid., 207.

43. Kahn, *Anaximander and the Origins of Greek Cosmology*, 97.

44. Retrospectively, Heidegger's 1950 essay 'Das Ding' seems to have spelt this out clearly: Heidegger proposed to understand the thing in terms of the fourfold: heaven, earth, divine, and mortal; it is not without relevance here to note that Reinhard May, in his *Ex Oriente Lux: Heideggers Werk unter Ostasiatischem Einfluss* (Wiesbaden: Franz Steiner Verlag, 1989) argues that the concept of emptiness (*Leere*) that Heidegger elaborated in 'Das Ding' comes from Chapter 11 of *Tao Te Ching*. If this claim is valid, then

§9. HARMONY AND THE HEAVEN

In contrast, in the absence of this conception of 'the uncanniest of the uncanny' of the human, the violence of *technē*, and the excessive violence of Being, in Chinese thought we find harmony—but we might also say that, for the Chinese, this fittingness resides in another kind of relation between humans and other cosmological beings, one which is based on *resonance* rather than war (*polemos*) and strife (*eris*). What is the nature of such resonance? In the *Classic of Poetry* (composed between the eleventh and seventh centuries BC), we can already find a brief description of the relation between the solar eclipse and the misconduct of King You of Zhou (周幽王, 781–771 BC).[45] In *Zuo Zhuan* (400 BC), a commentary on the ancient Chinese chronicle *Spring and Autumn Annals*, in the chapter on the Duke Yin, there is also a description of the relation between the solar eclipse and the death of the king.[46] In the *Huainanzi* (125 BC), a book reported to have been

---

Heidegger's 'move' to cosmotechnics becomes more evident. For a more developed account of the connection between Heidegger and Daoism, see M. Lin, *Heidegger on West-East Dialogue—Anticipating the Event* (New York and London: Routledge, 2008).

45.　'Minor odes of the kingdom (詩經·小雅·祈父之什·十月之交)': 'At the conjunction [of the sun and moon] in the tenth month / On the first day of the moon, which was Xin-mao / The sun was eclipsed / A thing of very evil omen. / Then the moon became small / And now the sun became small / Henceforth the lower people / Will be in a very deplorable case.' (十月之交、朔日辛卯。日有食之、亦孔之醜。彼月而微、此日而微。今此下民、亦孔之哀。). *Classic of Poetry*, tr. J. Legge, <http://ctext.org/book-of-poetry/decade-of-qi-fu>,

46.　'In his third year, in spring, in the king's second month, on the day Jisi, the sun was eclipsed. In the third month, on the day Gengxu, the king [by] Heaven's [grace] died.' (三年，春，王二月，己巳，日有食之。　三月，庚戌，天王崩。). *Zuo Zhuan* (左傳), 'The Third Year of Duke Yin' (隱公三年), <http://www2.iath.virginia.edu:8080/exist/cocoon/xwomen/texts/chunqiu/d2.7/1/0/bilingual>.

written by Liu An, the King of Huainan, and which attempts to define the socio-political order, we find many examples that depend upon the relation between the *Dao* of nature (as expressed in the Heaven) and the human. As various authors have explained, in ancient China heaven was understood both as an anthropomorphic heaven and as the heaven of nature. In the Confucian and Daoist teaching, the Heaven is not a deity; rather it is a moral being. The stars, winds, and other natural phenomena are indications of the reasons of the Heaven, which embodies objectivity and universality; and human activities must accord with these principles.

As we shall see, this conception of nature also inflects the thinking of time. Granet and Jullien both suggest that one should understand the expression of time in China not as linear or mechanical, but as seasonal, in the sense indicated by the changes of Heaven. In the following example from the chapter of the *Huainanzi* entitled 'Celestial Pattern', the different winds throughout the year are indicators of different political, social, and intellectual activities, including making sacrifices and executing criminals:

What are the eight winds?

Forty-five days after the winter solstice arrives, the Regular (northeast) Wind arrives.

Forty-five days after the Regular Wind arrives, the Brightly Abundant (east) Wind arrives.

Forty-five days after the Brightly Abundant Wind arrives, the Clear Bright (southeast) Wind arrives.

Forty-five days after the Clear Bright Wind arrives, the Sunshine (south) Wind arrives.

Forty-five days after the Sunshine Wind arrives, the Cooling (southwest) Wind arrives.

Forty-five days after the Cooling Wind arrives, the Changhe (west) Wind arrives.

Forty-five days after the Changhe wind arrives, the Buzhou (northwest) Wind arrives.

Forty-five days after the Buzhou wind arrives, the Broadly Expansive (north) wind arrives.

When the Regular Wind arrives, release those imprisoned for minor crimes and send away those (foreign intruders) who had been detained.

When the Brightly Abundant Wind arrives, rectify boundaries of fiefs and repair the fields.

When the Clear Bright Wind arrives, issue presents of silk cloth and send embassies to the Lords of the Land.

When the Sunshine Wind arrives, confer honors on men of position and reward the meritorious.

When the Cooling Wind arrives, report on the Potency of the earth and sacrifice at the four suburbs.

When the Changhe wind arrives, store away the suspended (bells) and hanging (chimestones); *qin* and *se* (stringed instruments) (must be) unstrung.

When the Buzhou wind arrives, repair palaces and dwellings and improve dikes and walls.

When the Broadly Extensive Wind arrives, close up gates and bridges and execute punishments.[47]

---

47. '何謂八風？距日冬至四十五日，條風至；條風至四十五日，明庶風至；明庶風至四十五日，清明風至；清明風至四十五日，景風至；景風至四十五日，涼風至；涼風至四十五日，閶闔風至；閶闔風至四十五日，不周風至；不周風至四十五日，廣漠至。條風至則出輕系，去稽留；明庶風至則正封疆，條田疇；清明風至則出幣帛，使諸候；景風至則爵有德，賞有功；涼風至則報地德，祀四郊；閶闔風至則收縣垂，琴瑟不張；不周風至則修宮室，繕邊城；廣漠風至則閉關梁，決刑罰。' *Huainanzi*, 3.12.

Indeed, what lies behind the whole discourse of the *Huainanzi*, as becomes more explicit in chapters such as 'Seasonal Rules' and 'Surveying Obscurities', is the concept of a resonance between human and Heaven which is real, not ideal or purely subjective, and which is also more than a question of signs or portents. This resonance is best demonstrated by the *Qin* and the *Se*, two musical instruments that produce a harmony with one another. For the Confucians, the resonance between human and Heaven is not purely subjective, but as objective and concrete as the resonance of these musical instruments.

The concept of resonance between the Heaven and humans was further elaborated in Han Confucianism, where it would be used as a legitimation of authority and of Confucianist teaching. In the same period as the *Huainanzi*, according to historians, both Daoism and Confucianism declined and became contaminated by certain superstitious modes of thought[48]—superstitious in the sense that these schools relied on supersensible mysterious powers which were sometimes incompatible with the Confucian teaching, for example in *Huang-Lao* (黃老)—a combination of Daoism and the Yin-Yang school which was on the verge of turning into a cult. It was in this context that Dong Zhongshu (董仲舒, 179–104 BC), the most important Confucian of the Han dynasty, employed the concept of the 'resonance between the Heaven and the human [天人感應]'.[49] Dong's contribution is a source of

---

48.   Lao Sze-Kwang (勞思光), *History of Chinese Philosophy—New Edition*, vol. 2 (《中國哲學史新編》第二冊) (Guilin: Guangxi Normal University Press), 11–24.

49.   According to Hu Shi (胡適, 1891–1962), *Outline of History of Chinese Philosophy* (中國哲學史大綱) (Shanghai: Shanghai Ancient Work Publishing House, 1997), Hu states that the concept of resonance between the heaven and the human was invented by Moism rather than Confucianism, although it was employed as the main theoretical tool by Confucianism during the Han

contention, since on the one hand he made Confucianism the principal doctrine of political thought and even of Chinese culture after him, which was to have a profound impact; on the other hand, many historians have criticised him for having introduced the superstitious thought of *Yin-Yang* and *Wu Xing* into Confucianism, and therefore of having transformed Confucianism from a discourse on human nature or *Xin Xing* (心性) into a discourse on the lawful Heaven which effectively gave authority to the emperor to carry out his political will.[50] Dong's approach to understanding the relation between the Heaven and the moral order is nonetheless similar to that found in the *Huainanzi*. *Yin* and *Yang* are understood respectively as moral good and punishment, corresponding to summer and winter. Even though most historians have agreed that Dong's interpretation is not authentically Confucian, and conceded that his theory was in the service of feudalism, it is crucial to recognise that this relation that he envisioned between the human and the Heaven did not come from nowhere; it is already hinted at in the early classics such as Zhuangzi and Laozi, where it is authorised by a moral-cosmological view of nature—namely, the unification between the human and the Heaven (天人合一). We can understand this point better when reading Dong's suggestions to the emperor:

If the emperor wants to achieve something, it is better to ask the Heaven. The way of Heaven is based on *Yin-Yang*. *Yang* is virtue [*de*], *Yin* is punishment. Punishment corresponds to killing,

dynasty; Xu Dishan (1893–1941, 許地山) in *History of Daoism* (道教史) (Hong Kong: Open Page Publishing, 2012), 288, further points out that Daoism also adopted this concept.

50.   Ibid., 16. Lao Szekwang (1927–2012) argues that the deterioration of Han Confucianism is undeniable.

virtue corresponds to righteousness [*yi*]. Therefore, *Yang* dwells in summer, and occupies itself with growing; *Yin* dwells in winter, and accumulates in the void [...].

Unlike the early Greek thinkers, then, who sought to understand the question of *dikē* through the confrontation between human and nature, as described by Heidegger, and unlike the Greek rulers, who sought to impose *dikē* in order to overcome the excessiveness of the human, a spirit that we find the ancient Greek tragedies, the ancient Chinese seem to have endowed the cosmos with a profound morality expressed as a harmony which political and social life must follow, with the emperor as the intermediary between the Heaven and his people: he must cultivate his virtue by studying the classics and through constant self-reflection (by way of resonance with others), in order to put things in their proper order, convenient both to Heaven and to his people:[51]

> I heard that Heaven is the origin of all beings [...] so the sages
> follow Heaven in order to establish the way [*Dao*], therefore they
> have love for all and don't take any standpoint from their own
> interest [...] Spring is the vibrant moment of Heaven, when the
> Emperor will spread his benevolence; summer is the growing
> moment of Heaven, when the Emperor cultivates his virtue
> [*de*]; winter is the destructive moment of Heaven, when the
> Emperor executes his punishments. Therefore, the resonance

---

51. This point is valid not only for Confucianism, but also for Daoism, as is stated clearly in the *Zhuangzi* (see 'Heaven and Earth'); not to mention that the *Dao De Ching* (Laozi) is understood to be a guide for the emperor (帝王術).

between Heaven and human is the way [*Dao*], from the ancient times to the present.[52]

Cosmological Confucianism declined towards the end of the Han dynasty (206 BC–220 CE) for various reasons, the most significant of which was natural catastrophe. An equivalence between cosmological order and moral order means that cosmological disorder immediately implies moral disorder, and numerous natural disasters occurred during this period. Worse still, it was the period during which sunspots happened very frequently. All of this had the effect of calling cosmological Confucianism into question and destroying its credibility. As the historians Jin Guantao and Liu Chingfeng have pointed out, the failure of cosmological Confucianism led to the adoption, as a replacement, of the thinking of Laozi and Zhuangzi on nature and freedom, which emphasises 'non-doing', 'non-intervening'.[53] This is what is known as *Wei Jin Xuan Xue* (魏晉玄學), where *Xuan Xue*, literally meaning 'mysterious learning', is a term used to describe a form of thinking that lies somewhere between metaphysics in the Western sense and superstition. For this reason, some historians of philosophy have regarded the thinking that emerged during this period as superficial; later (§16.1) we will see how the term *Xuan Xue* was used to discredit Chinese intellectuals who had embraced the thought of Henri Bergson and Rudolf Eucken. At this point,

---

52. Lao Sze-Kwang, *History of Chinese Philosophy*., 27, 「然則王者欲有所為，宜求其端於天。天道之大者在陰陽。陽為德，陰為刑；刑主殺而德主生。是故陽常居大夏，而以生育養長為事；陰常居大冬，而積於空虛不用之處……臣聞，天者，群物之祖也。……故聖人法天而立道，亦溥愛而亡私。……春者，天之所以生也；仁者，君之所愛也；夏者，天之所以長也；德者，君之所以養也；霜者，天之所以殺也；刑者，君之所以罰也。由此言之，天人之征，古今之道也。」

53. Jiang Guantao (金觀濤) and Liu Chingfeng (劉青峰), *Ten Lectures on the History of Chinese Thought* (中國思想史十講) (Beijing: Law Press, 2015), 126.

though, we wish to emphasise that, although cosmological Confucianism may have declined, the importance of the relation between the Heaven and morality was preserved. As French physiocrat François Quesnay remarked in his 1767 essay *Despotism of China*, following a natural disaster in 1725 the Chinese emperor pleaded to the Heaven that it was his fault and not the people's; since it was his virtue that had been proved 'insufficient', he should be the one to be punished.[54] Indeed, this form of governance is still present today, as witnessed by the tears and speeches of the chairman or prime minister upon visiting the sites of natural or industrial disasters—for example, during the 2008 earthquake in Sichuan, when the prime minister Wen Jiabao visited the site, and his tears were the focus of media attention.

Notwithstanding the fierce critique against Dong's assimilation of Daoism and *Yin-Yang* into Confucianism, which was seen as a corruption of the 'pure' Confucian teaching, the unity between the cosmos and the moral has continued to be affirmed throughout the history of Chinese philosophy. This correlation between natural phenomena and the conduct of the emperor or the rise and fall of the empire may seem superstitious to us, yet it is worth emphasizing that the underlying spirit of such gestures, which continues after Dong, goes far beyond the mere correlation one might imagine, for example in mapping the number of solar eclipses and disasters in the Empire. The identification of moral with cosmic order draws its legitimacy not merely from the accuracy of such correlation, but rather from the belief that there is a unity between the Heaven and the human, which can be conceived of as a kind of

---

54. See F. Quesnay, *Œuvres économiques et philosophiques de François Quesnay Fondateur du système physiocratique* (Paris: Peelman, 1888), 563–660.

auto-affection.[55] It implies an inseparability of the cosmos and the moral in Chinese philosophy. On this point it is enlightening to turn to Mou Zongsan's critique of Dong. In *Nineteen Lectures on Chinese Philosophy*, Mou denounced Dong's thought as a cosmocentrism, since for Dong, the cosmos is prior to the moral, and therefore the cosmos becomes the explanation of the moral.[56] Mou's critique is no doubt justified; yet is it any more logical to place the moral prior to the cosmos? The moral can be established only when the human is already in-the-world, and being in-the-world only gains its profound meaning in the presence of a cosmology or principles of heaven—otherwise it would be only something like the animal-*Umwelt* relation described by Jakob von Uexküll. A few pages later, Mou also affirmed that, in *Doctrine of the Mean* (中庸) and *Yi Zhuang* (易傳), 'cosmic order is moral order'.[57] In Mou's interpretation of the whole tradition of Neo-Confucianism, then, this unity of the cosmic order and the moral order is always central, although, as we shall see (§18), because of his affinity to the work of Kant, for Mou *xin* ('heart') is posited as the absolute beginning. What we wish to emphasise here is that the unity between the cosmos and the moral is characteristic of ancient Chinese philosophy, and that this unity was further developed in the Neo-Confucianism that emerged from the time of the late Tang dynasty.

---

55.   Of course, this presumes the legitimacy of the Emperor; here we abstract from this context in order to address the cosmos and moral unity as an ontological question.

56.   Mou Zongsan, *Nineteen Lectures on Chinese Philosophy* (中國哲學十九 講) (Shanghai: Ancient Works Publishing House, 2005), 61.

57.   Ibid., 65.

§10. DAO AND QI: VIRTUE CONTRA FREEDOM

In Chinese thinking, *Dao* is superior to any technical and instrumental thinking, and the goal of *Dao* is also to transcend the limitations of technical objects—that is, to let them be guided by *Dao*. In contrast, it seems that the ancient Greeks, had a rather instrumental concept of *technē* as a means to an end, at least this was the case for the Aristotelians. The case of Plato is more complex. Whether *technē* plays a role in moral and ethical life in the dialogues of Plato is still a subject of debate among classical scholars. *Technē* is thought to derive from the Indo-European root *tek*, meaning 'to fit together the woodwork of [...] a house'.[58] For the Presocratics, the meaning of *technē* is closest to this root, and as Heidegger says, 'each *technē* is correlated with a quite determinate [*bestimmte*] task and type of achievement'.[59] Jörg Kube notes that, in Homer, the word *technē* is only used in relation to the god Hephaestus, or to carpentry, but not any other work, probably because other practices such as medicine, fortune-telling, and music had yet to become independent professions.[60] In Plato, we see a significant modification of the sense of the word, and it becomes closely related to another word, *aretē*, meaning

58.   T. Angier, *Technē in Aristotle's Ethics. Crafting the Moral Life* (London and New York: Continuum, 2012), 3.

59.   F. Heinimann, 'Eine Vorplatonische Theorie der τεχνη', *Museum Helveticum* 18:3 (1961), 106; cited by Angier, *Technē in Aristotle's Ethics*, 3.

60.   J. Kube, *TEXNH und APETH: Sophistisches und Platonisches Tugendwissen* (Berlin: De Gruyter, 1969), 14–15. In The Iliad, Paris compares Hector's heart with the axe of a carpenter who uses 'technē' to cut a beam to be used in a ship; Roochnik (*Of Art and Wisdom*, 23) further pointed out that in Odysseus, one can find two words derived from *technē*, *technēssai* (the Phaeacian woman skilled at weaving) and *technēentēs* (Odysseus's skillful steering of a ship).

'excellence' in general, 'virtue' in particular.[61] Vernant remarks that the word *aretē* had started undergoing a shift already in the time of Solon (640–558 BC), where its relation to the warrior in the aristocratic milieu had been transferred to another conception of self-control belonging to the religious milieu: correct behaviour resulting from a long and painful *askesis*, and which aims to resist *koros* (greed), *hybris* (excess) and *pleonexia* (avarice), the three follies. The 'human cosmos' (the *polis*) is conceived to be a harmonious unity in which the individual *aretē* is *sōphrosynē* (temperance), and *dikē* is a law common to all.[62] As Vernant says, 'with Solon, *dikē* and *sōphrosynē* descend from heaven to earth, to be installed in the agora'.[63] Virtue-*technē* constitutes a core enquiry in Plato's quest for a *technē* of all *technai* that can be learned and taught, and for *dikē* as the virtue of all virtues.[64] Each *technē* is a remedy for overcoming chance occurrences (*tychē*) and errors that crop up in the process of making, as Antiphon says: 'we conquer by *technē* things that defeat us by *physis*'.[65] This motif is repeated many times in Plato's dialogues. Notably, in *Protagoras* Socrates admires the figure of Prometheus, and agrees with Protagoras in affirming the necessity of measurement (*metrētikē technē*) as a way to restrict hedonism, as

61. L. Brisson, '*Tekhnê* is Not Productive Craft', Preface to A. Balansard, *Technè dans les dialogues de Platon* (Sankt Augustin: Academia Verlag, 2001), XI.

62. J-P. Vernant, *Les origines de la pensée grecque* (Paris: PUF, 1962), 92–3.

63. Ibid., 96.

64. Zore, 'Platonic Understanding of Justice. On *dikē* and *dikaiosyne* in Greek Philosophy', in Barbarić (ed.), *Plato on Goodness and Justice*, 29.

65. Angier, *Technē in Aristotle's Ethics*, 4.

well as the elimination of *tychē*.[66] In *Gorgias*, Polus claims that 'experience causes our time to march along the way of *technē*, whereas inexperience causes it to march along the way of *tychē*'.[67] The relation between the cosmos (order) and geometry is clearer in the later passages in the *Gorgias* when Socrates tells Callicles that, according to the wise man who had made a study of geometry,

> partnership and friendship, orderliness, self-control, and justice
> hold together heaven and earth, and gods and men, and that is
> why they call this universe a world order [...] You have failed to
> notice that proportionate equality has great power among both
> gods and men, and you suppose that you ought to practice get-
> ting the greater share. That is because you neglect geometry.[68]

In the *Timaeus*, too, the universe is 'a work of craft (*dedēmiourgētai*) grasped by a rational account (*logōi*)—that is, by wisdom (*phronēsis*)'.[69] This is precisely because what Plato continually seeks is a *technē* of justice (*dikē*, *dikaiosynē*), a reason that is correct for the self and for the community; a *technē* which, in this regard, would not simply be one technique among others, but the technique of all *technai*.

---

66. In *The Fragility of Goodness* Nussbaum claims that this desire to eliminate chance (*tychē*) led to the decline of Greek tragedy; an argument that resonates with Nietzsche's *Birth of Tragedy*, where the figure of Socrates, who introduced reason as an Apollonian measurement, precipitated the decline of the Dionysian spirit.

67. Plato, 'Gorgias', in *Complete Works*, 448c.

68. Ibid., 508a.

69. Plato, 'Timaeus', in *Complete Works*, 29a; cited by Angier, *Technē in Aristotle's Ethics*, 18–19.

For the ancient Greeks, then, *technē* implies a *poiētikē* that brings about a proper end, a good end. In Plato's *Phaedrus* we find a distinction between *technai* and *technēmata*, where the latter means simply 'techniques': a doctor who can cure, for example by lowering or raising the body temperature, has mastered *technē*, but one who only knows how to bring about insignificant or negative changes in the patient 'knows nothing of the *technē*'.[70] *Technē*, which aims for the good, is not immediately given, being neither an innate gift nor one endowed by a divine power (like poetry), but is rather something that needs to be mastered. Notably, in Book II of the *Republic* (374d–e), Socrates tells us that

> No [...] tool makes anyone who picks it up a *dēmiourgos*, unless he has acquired the requisite knowledge and has had sufficient practice [...]. Then to the degree that the *ergon* of the guardians is most important, [...] it requires [...] the greatest *technē*.[71]

A few more words are necessary on the relation between *technē* and *aretē*, since it is important for the interpretation of Plato and Aristotle, and hence also gives us a comprehensive understanding of the technical question handed down from classical Greek philosophy. The relation between the two remains an important point of contention among classical scholars. I do not aim to enter into this debate here, but only to give an overview of it. Indeed, it may in fact be more interesting to start with the question: What is *not technē*? Vernant differentiates *technē* from *praxis*, a distinction that arguably follows from the logic of Critias's challenge to Socrates in the

---

70. Plato, 'Phaedrus', in *Complete Works*, 268c.

71. Cited by Angier, *Technē in Aristotle's Ethics*, 31.

Charmides, where he states that *technē* as *poeisis* always has a product (*ergon*), while *praxis* has its end in itself.[72] This is rather debatable, however; indeed, it also indicates the complexity of Plato's concept of *technē*. For example, the sophists also have *technē*, yet it is not a *technē* of production (*poiētikē*), but a *technē* of acquisition (*ktētikē*).[73] The other thing that is opposed to *technē* is *empeiria*, often translated as 'experience', since it is said to be subject to illusion and error. Poetry is also not *technē*, but in a different way, since a good poet is not the real author, but a channel for a divine power (*thēia moira*).[74] Hence we can see that, as Nussbaum has shown, a common point in distinguishing *technai* from non-technics is that the objective of *technē* is to overcome *tychē*, to become the guarantor of order, of proportion, like the *Demiourgos* in the *Timaeus*. How is it related to virtue, then? For simplicity, I would summarise the relation between *technē* and *aretē* in the following ways:

Technē as analogy of aretē. In different dialogues, Socrates tries to draw analogies between *technē* and *aretē*: courage in the *Laches*, temperance in the *Charmides*, piety in the *Euthyphro*, justice in the *Republic*, wisdom in the *Euthydemus*.[75] But in the *Charmides*, Critias challenges Socrates for comparing temperance (*sōphrosynē*) with other *technai* such as medicine or masonry because, like calculation and geometry,

---

72. Balansard, *Technè dans les dialogues de Platon*, 6.

73. Ibid., 78

74. Ibid., 119.

75. Roochnik, *Of Art and Wisdom*, 89–177.

temperance has no product (*ergon*), while medicine aims for health and masonry aims for a house.[76]

Aretē as the aim of technē: This point is not immediately evident, since although on many occasions Socrates uses medicine as an example of *technē*, in other cases *technē* is considered to be neutral (not necessarily good or bad). However, a passage in the *Gorgias* seems to reveal this point in a striking manner: Socrates replies to Polos that cooking is not a *technē*, but that a knowledge of cookery is just knowing how to favour and please.[77] The reason given is that cooking is a 'forgery of medicine' because 'it pursues pleasure but not the health of the body'.[78]

Aretē as technē: David Roochnik claims that this relation becomes evident in the middle period of Plato's writing, for example in Books II–X of the *Republic*,[79] where justice is considered to be a philosophical *technē*, a judgement of proportion, as invoked at the beginning of the *Timaeus*—as in myth, where, in view of the incompleteness of the *technai* brought to man by Prometheus, Zeus sent respect (*aidôs*) and justice (*dikē*) to human being as *politikē technē*.[80]

This *technē-aretē* relation breaks down in Aristotle's classification of knowledge in Book VI of the *Nicomachean Ethics*. In Plato's time, as some philologists have argued, there

---

76. Plato, 'Charmides', in *Complete Works*, 165e3–166a1.

77. Plato, 'Gorgias', in *Complete Works*, 462d8–e1.

78. Balansard, *Technè dans les dialogues de Platon*, 141.

79. Roochnik, *Of Art and Wisdom*, 133.

80. Balansard, *Technè dans les dialogues de Platon*, 93.

was no systematic or general distinction between *epistēmē* and *technē*: since *technē* doesn't necessarily have an *ergon*, *epistēmē* can in some cases be counted as a form of *technē*.[81] In contrast, in Aristotle's *Nicomachean Ethics technē* is strictly distinguished from *epistēmē*, which concerns knowledge of the unchangeable; it is also distinguished from *phronesis*, practical wisdom, for a familiar reason that we have already encountered: *technē* has a product, while *praxis* has none. Aristotle affirms this distinction: 'making [*poiēton*] and acting [*praktikon*] are different [...] hence too they are not included in each other [...]'.[82] *Technē*, often translated as 'art' in this context, is a form of production in which something is accomplished against all other possibilities, meaning against chance. Aristotle cites Agathon's remark that 'art loves chance and chance loves art'.[83] We should point out here that, no matter whether it is regarded as a process of making or a form of praxis, *technē* is considered as an assurance, as a means toward excellence and virtue, as the word *aretē* signifies. The two hundred pages of commentary Heidegger dedicates to Book VI of the *Nicomachean Ethics* in his 1924–5 lecture course published as *Plato's Sophist*, however, disturbs this neat classification. *Technē*, he affirms, following Plato, is not 'producing' or 'making'

---

81.   Nussbaum, *The Fragility of Goodness*, 94: 'to judge from my own work and in the consensus of philologists, there is, at least through Plato's time, no systematic or general distinction between *epistēmē* and *technē*. Even in some of Aristotle's most important writings on this topic, the two terms are used interchangeably.'; Heidegger, 'The Question Concerning Technology', 13: 'from earliest times until Plato the word *technē* is linked with the word *epistēmē*. Both words are names for knowing in the widest sense. They mean to be entirely at home in something, to understand and be expert in it. Such knowing provides an opening up. As an opening up it is a revealing.'

82.   Aristotle, 'Nicomachean Ethics', in *The Basic Works of Aristotle* (New York: Modern Library, 2001), 1025 (1140a).

83.   Ibid., 1140a20.

but rather 'seeing', 'grasping the essence of' things in question or things to come. As Heidegger says,

> The one who has *technē* is admired, even if he lacks the practical skill of the hand-laborers, precisely because he sees the essence. He may thereby fail in practice, for practice concerns the particular, whereas *technē* concerns the universal. Despite the shortcoming with regard to practice, the one who has *technē* is still respected more and considered wiser: in virtue of his privileged way of looking disclosively.[84]

Furthermore, Heidegger points out that in the same passages of Book VI, *sophia* is designated as the excellence (*aretē*) of *technē*,[85] and that philosophy is nothing more than the pursuit of this excellence. Here Aristotle's classification is not strictly followed—instead, Heidegger returns to the *technē* of Plato. Heidegger blends Plato and Aristotle together here, but only in order to point out that, no matter whether it is regarded as a process of making or a form of *praxis*, *technē* is considered as an assurance, as a means toward excellence and virtue, as the word *aretē* signifies.

Following this brief outline of the concept of *technē* in Plato and in Aristotle, we must now come to Heidegger's reading of their metaphysics as declension (*Abfall*) and fall (*Absturz*).[86] If early Greek thinkers such as Parmenides, Heraclitus, and Anaximander are what Heidegger calls inceptual (*anfänglicher*) thinkers, in the sense that they think about the

---

84. Heidegger, *Plato's Sophist* (Bloomington, IN: Indiana University Press, 1997), 52; cited by R. Rojcewicz, *The Gods and Technology: A Reading of Heidegger* (Albany, NY: State University of New York Press, 2006), 63–4.

85. Heidegger, *Plato's Sophist*, 39

86. Boehm, 'Pensée et technique', 202.

beginning rather than presence, and if for them there is no clear distinction between Being and beings, in Plato and Aristotle Heidegger finds a passage from pre-metaphysics to metaphysics proper, a passage which shaped the history of metaphysics as history of ontotheology. It is this metaphysics, begun by Plato and Aristotle and completed in Hegel and Nietzsche,[87] which finally leads to *Gestell* as the essence of modern technology. American Heidegger scholar Michael Zimmerman calls it 'productionist metaphysics',[88] because such a metaphysics is concerned with production or the technical from its very beginning, ending up with 'machination' (*Machenschaft*) and *Gestell*. Ontotheology bears with it two questions: firstly, what are beings as such (ontology)? Secondly, what is the highest being (theology)? The Idea of the Good (*hē tou agathou idea*) in Plato sets out such an ontotheological beginning, since it is that which 'makes intelligible things intelligible', and provides 'truth/disclosure to what is known and endows the one who knows with a capacity to know'.[89] It signifies a determination of the essence (*ousia*) by subsuming the many to the one, the Idea; and in this sense, the 'Idea' is also the 'good' since it is the cause for all, which Aristotle calls *to theion*, the divine:[90]

---

87. Backman, *Complicated Presence*, 13.

88. M.E. Zimmerman, *Heidegger's Confrontation with Modernity: Technology, Politics, and Art* (Indianapolis: Indiana University Press, 1990), 3, 'He [Heidegger] believed that the Greeks initiated "productionist metaphysics" when they concluded that for an entity "to be" meant for it to be produced. While what they meant by "production" and "making," for Heidegger, differed from the production processes involved in industrial technology, still it was the Greek understanding of the being of entities which eventually led to modern technology.'

89. Backman, *Complicated Presence*, 37.

90. M. Heidegger, *GA 9, Wegmarken* (Frankfurt am Main: Klostermann, 1996), 235.

Ever since being has been explicated as idea, the thinking of the being of beings has been metaphysical, and metaphysics theological. 'Theology' means in this case the explication of the 'cause' of beings as God and the relocation of being into this cause, which contains being in itself and also releases being from out of itself, because it is the most beingful [*Seiendste*] of beings.[91]

Ontotheology continued to develop in Neoplatonic metaphysics and Christian philosophy, finally leading to the oblivion of Being and the abandonment of Being—the age of the *Bestand*.[92] This history of ontotheology and productionist metaphysics was apparently absent in China, and indeed we find a very different relation between technics and virtue in Chinese cosmotechnics, a different form of 'belonging together' or 'gathering' than that to which Heidegger aspired, one based on an organic form guided by a moral and cosmological consciousness. But before explaining this concept in more detail, I would like to go back to the relation between the two fundamental categories *Qi* and *Dao* in Chinese philosophy. We have said that *Qi* (器) means 'tool', but in fact there are three different words which are not clearly distinguished in everyday (especially modern) Chinese:

Ji (機)—That which controls the trigger (主發謂之機。从木幾聲。).
Qi (器)—Containers being guarded by dogs (皿也。象器之口，犬所以守之。).
Xie (械)—Shackle, also called *Qi*, meaning 'holding'; one says

---

91. Ibid., 235–6; cited by Backman, *Complicated Presence*, 43–4.
92. Backman, *Complicated Presence*, 55.

that which contains is *Xie*, and that which doesn't contain is *Qi* (桎梏也。从木戒聲。一曰器之總名。一曰持也。一曰有盛爲械，無盛爲器。).

Two different compound words, 機器 (*Jī Qì*) and 機械 (*Jī Xiè*), refer to machines, and the words are interchangeable. Even in the dictionary of ancient etymology (說文解字) there is ambiguity on this point: for example, in the entry on *Qi*, we are told that it is a container; however, in the entry on *Xie*, *Qi* is said to be the synonym of *Xie*. From the pictogram of *Qi*—four mouths or openings with a dog in the middle—we can see that *Qi* implies a virtual spatial form, while *Xie*, with the pictogram for wood on the left, refers to actual material tools, and is also closely related to instruments of torture. *Qi* has the pictogram of four squares surrounding a dog, as if the dog is surveilling the space and guarding the containers. The four squares can also be the character for 'mouth', and hence are related to living (drinking, eating). *Ji* is more straightforward to understand, since it has the clearest meaning of machinery: that of triggering something else and setting it in motion.

The spatial form of *Qi* is technical, though, in the sense that it imposes forms. In the *Xi Ci*, a commentary on the *I Ching*, we read that 'what is formless (or above form) is called *Dao*; what has form (or is below form) is *Qi* (形而上者謂之道，形而下者謂之器)'. In the same text, we read that 'if there is appearance, then we call it phenomena; if there is form, we call it *Qi* (如見乃謂之象，如形乃謂之器)'. It is important to note that *xing er shang* (形而上, 'above form'), is used to translate the English word 'metaphysical'; and *xing er shang xue* (形而上學) is the study of it—metaphysics. *Dao* is what gives form and phenomenon; it is what is above them, as superior being.

However, *Dao* does not mean the 'laws of nature' as this term was understood in seventeenth-century Europe; it is rather the ungraspable that could yet be known. The commentator Zheng Xuan (鄭玄, 127–200), combining his reading of Laozi with the *I Ching*, says that 'the universe doesn't have form in its origin. Now we find forms, because forms come from the formless. That is why the *Xi Ci* says "what is above form is called *Dao*"'.[93] *Qi* is also a container, a carrier, but not only a physical one; it also means specificity and generosity. In *The Analects of Confucius*, one reads '*junzi bu qi* (君子不器)', where *junzi* is the ideal personality of Confucians. This phrase is often translated as 'the gentleman is not a utensil',[94] meaning that he doesn't limit himself to any specific purpose; it could also be read as saying that his generosity is unbounded. In this sense, *Qi* is the bounded, finite being that emerges according to the infinite *Dao*.

It is in attending to the relation between *Dao* and *Qi* that we can reformulate a philosophy of technology in China. Now, this relation has a subtle similarity to the *technē-aretē* relation discussed above—but is also very different in the sense that it exhibits another, rather different cosmotechnics, one which searches for a harmony based on the organic exchanges between the cosmos and the moral. Chinese philosopher of technology Li Sanhu's excellent *Reiterating Tradition: A Comparative Study on The Holistic Philosophy of Technology*,[95]

93.   Wu Shufei (吳述霏), 'Analysis of "Below and Above Form" in the *I Ching*' (周易「形而上、下」命題解析), *Renwen* (《人文》) 150 (June, 2006). '天地本無形，而得有形，則有形生於無形矣。故《系辭》曰：『形而上者謂之道』'.

94.   Confucius, *The Analects of Confucius*, tr. B. Watson (New York: Columbia University Press, 2007), 21.

95.   Li Sanhu (李三虎), *Reiterating Tradition: A Comparative Study on The Holistic Philosophy of Technology* (重申傳統：一種整體論的比較技術哲學研究) (Beijing: China Social Science Press, 2008).

which can without exaggeration be called the first attempt to seek genuine communication between technological thought in China and the West, calls for a return to the discourse on *Qi* and *Dao*. Li tries to show that *Qi*, in its original (topological and spatial) sense, is an opening to the *Dao*. Hence Chinese technical thought comprises a holistic view in which *Qi* and *Dao* reunite to become One (道器合一). Thus the two basic philosophical categories of *Dao* and *Qi* are inseparable: *Dao* needs *Qi* to carry it in order to be manifested in sensible forms; *Qi* needs *Dao* in order to become perfect (in Daoism) or sacred (in Confucianism), since *Dao* operates a privation of the determination of *Qi*.

§10.1. QI AND DAO IN DAOISM: PAO DING'S KNIFE

Whereas for Plato virtue-technics is fundamentally a matter of measurement, an exercise of reason in its quest for a form that allows self-governance and the governance of the *polis*, in the *Zhuangzi*, *Dao* as technics is the ultimate knowledge devoid of measurement, since it is *zi ran*. The Daoists aspire to *zi ran*, and see *zi ran* as the most superior knowledge, namely that of *wu wei* (無為, 'non-doing'). Accordingly, the Daoist principle of governance is *wu wei zhi zhi* (無為之治), which means governing *without* intervening. This is no pessimism or passivism, but rather a letting things be, leaving room for things to grow on their own, in the hope that beings will fully realise themselves and their potential—a point that is emphasised by Guo Xiang (郭象, 252–312) in his commentary on the *Zhuangzi*.[96] Unlike Wang Bi, the commentator of the

---

96. Cited by Lao Szekwang, *History of Chinese Philosophy* vol. 2, 146. 「無為者，非拱默之謂也，直各任其自為，則性命安矣。」.

I Ching and *Dao De Ching* during the same period of Wei Jin, who believes that *wu* (無, nothing) is the foundation of *Dao*, and those after Wang Bi who believe that the foundation should be *you* (有, 'being', 'there is'), Guo Xiang criticised such an opposition as futile, since being cannot come from nothing, and one being does not yield all beings; instead, he proposes to understand the foundation of *Dao* in terms of *zi ran*—following the principle of the cosmos without unnecessary intervention.[97]

In order to better understand the essence of Daoist cosmotechnics, we might refer here to the story of the butcher Pao Ding, as told in the *Zhuangzi*. Pao Ding is excellent at dissecting cows, but according to him, the key to being a good butcher doesn't lie in his mastery of the skill, but rather in comprehending the *Dao*. Replying to a question from the prince Wen Huei about the *Dao* of butchering a cow, Pao Ding points out that having a good knife is not necessarily enough; it is more important to understand the *Dao* in the cow, so that one does not use the blade to confront the bones and tendons, but rather passes alongside them in order to enter into the gaps between them. Here the literal meaning of '*Dao*'—'way' or 'path'—meshes with its metaphysical sense:

> What I love is *Dao*, which is much more splendid than my skill. When I first began to carve a bullock, I saw nothing but the whole bullock. Three years later, I no longer saw the bullock as a whole but in parts. Now I work on it by intuition and do not look at it with my eyes. My visual organs stop functioning while my intuition goes its own way. In accordance with the principle

---

97. See Jin Guantao and Liu Chingfeng, *Ten Lectures*, 149 「無既無矣,則不能生有」,「豈有之所能有乎?」.

of heaven (nature), I cleave along the main seams and thrust the knife into the big cavities. Following the natural structure of the bullock, I never touch veins or tendons, much less the big bones![98]

Hence, Pao Ding concludes, a good butcher doesn't rely on the technical objects at his disposal, but rather on *Dao*, since *Dao* is more fundamental than the *Qi* (tool). Pao Ding adds that a good butcher has to change his knife once a year because he cuts through tendons; a bad butcher changes his knife every month, because he directly chops the bones with the knife; while Pao Ding has not changed his knife for nineteen years, and yet it looks as if it had just been sharpened with a whetstone. Whenever Pao Ding encounters any difficulty, he slows down the knife, and gropes for the right place to move further.

The prince Wen Huei, who had posed the question, replies that 'having heard from Pao Ding, now I know how to *live*'; and indeed this story is included in the section titled 'Master of Living'. Moreover, it is the question of 'living', rather than that of technics, that is at the centre of the story. If there is a concept of 'technics' here, it is one that is detached from the technical object: although the technical object is not without importance, one cannot seek the perfection of technics through the perfection of a tool or a skill, since perfection can only be accomplished by *Dao*. When it is used in accordance with instrumental reason—with functions such as 'chopping', and 'cutting'—the knife performs actions that belong only to the lower level of its being. When the knife is guided by *Dao*, on the

---

98. *Zhuangzi* (bilingual edition) (Hunan: Hunan People's Publishing House, 2004), 44–5 [translation modified].

other hand, it becomes perfect through the 'privation' of the functional determinations imposed upon it by the blacksmith. Every tool is subject to technical and social determination which endow it with its specialised functions—for example, the kitchen knife has a technical determination with a sharp blade, and a social determination for culinary use. 'Privation' here means that Pao Ding does not exploit the purposely built-in features of the knife—being sharp for cutting and chopping—but endows it with a new usage in order to fully realise its potential (as being sharp). Pao Ding's knife never cuts the tendons, not to mention coming up against the bones: instead, it seeks the void and enters it with ease; in doing so, the knife accomplishes the task of butchering the cow without endangering itself, i.e. becoming blunt and then needing to be replaced, and fully realizes itself as a knife.

The knowledge of living thus consists of two parts: understanding a general principle of life, and becoming free from functional determination. This could be regarded as one of the highest principles of Chinese thinking on technics. However, we must also note that *Dao* is not only the *principle of being*, but also the *freedom to be*. In this particular conception of *Dao*, then, *Dao* may not lead technics to its perfection; indeed, *Dao* may be subverted or even perverted by technics. We find this concern in a story in the section of the *Zhuangzi* entitled 'Heaven and Earth', in which the character Zigong (who shares this name with one of Confucius's most famous students, known as a businessman) encounters an old man who is occupied with manually carrying water from a well to his farm. Having seen that the old man has 'used up a great deal of energy and produced very little result', Zigong intervenes:

'There is a machine for this sort of thing', said Zigong. 'In one day it can carry water across a hundred fields, demanding very little effort and producing excellent results. Wouldn't you like one?'

The gardener raised his head and looked at Zigong. 'How does it work?'

'It's a contraption made by shaping a piece of wood. The back end is heavy and the front end light, and it raises the water as though it were pouring it out, so fast that it seems to boil right over! It's called a well sweep.'

The gardener flushed with anger and then said with a laugh, 'I've heard my teacher say, where there are machines, there are bound to be machine worries; where there are machine worries, there are bound to be machine hearts [ji xin, 機心]. With a machine heart in your breast, you've spoiled what was pure and simple, and without the pure and simple, the life of the spirit knows no rest. Where the life of the spirit knows no rest, the Way [Dao] will cease to buoy you up. It's not that I don't know about your machine—I would be ashamed to use it!'

Zigong blushed with chagrin, looked down, and made no reply. After a while, the gardener said, 'Who are you, anyway?'

'A disciple of Kong Qiu [Confucius].'[99]

Considering that this is a dramatic encounter between Confucius's student and Zhuangzi's student, we can read it as a mockery of Confucius, who was busy with his political activities, meaning that he himself might be considered one 'who has spoiled what was pure and simple'. Here machines are conceived to be tricks; devices that deviate the simple and pure into complications that inevitably spoil a form of life. Machines demand a form of reasoning that causes Dao to

99. Zhuangzi, *The Complete Works of Zhuangzi*, 90–91.

deviate from its pure form, which in turn gives rise to anxiety. *Ji xin* (機心), however, is not best rendered as 'machine heart', as the translation suggests, but rather as 'calculative mind'. The old man affirms that he is aware of the existence of this machine, and that it was also known to his teacher, but that they felt ashamed to use it and so refused this technique. What Zhuangzi wants to say in this story is that one should avoid developing such a reasoning about life, otherwise one will lose the way, and along with it, one's freedom; if one always thinks in terms of machines, one will develop a machinic form of reasoning.

To conclude this section, we must acknowledge that in the *Phaedrus*, there are two passages that greatly resemble these two stories from the *Zhuangzi*, yet which also exhibit significant differences. Plato evokes an art that may seem similar to Pao Ding's skill in butchery: after having delivered two speeches against Lysias's argument that 'it is better to give your favour to someone who does not love you than to someone who does',[100] Socrates comments on the art of rhetoric. He explains to Phaedrus that there are two forms of 'systematic art':

> The first consists in seeing together things that are scattered about everywhere and collecting them into one kind [...] [the other] in turn, is to be able to cut up each kind according to its species along its natural joints, and to try not to splinter any part, as a butcher might do.[101]

---

100. Plato, 'Phaedrus', in *Complete Works*, 227c.
101. Ibid., 265d–e.

Socrates here places emphasis on the need to know the nature of things, as a medical doctor knows the nature of the body, and a rhetorician knows the nature of the soul. A rhetorician, by knowing the soul, is able to direct souls according to their different types, by choosing different words. For Plato, arts such as rhetoric and medicine must know the nature of things, otherwise they risk being mere 'empirical and artless practice[s]'.[102] Zhuangzi's stories, on the other hand, are more concerned with a way of living; to live well is not to confront the 'hard' and the 'extreme', for instance by adopting the impossible task of pursuing infinite knowledge in one's finite life, but to learn how to live by following *Dao*, which, as he clearly insists, for the butcher is not a mere matter of knowing anatomy.

The second episode in Plato is the famous story told by Socrates of the Egyptian god Theuth, the inventor of number, calculation, geometry, astronomy, and writing. Theuth comes to the King of Egypt Thamus and exhibits his arts. When it comes to writing, the King disagrees with Theuth, protesting that writing actually has the opposite effect to that which Theuth ascribes to it: rather than aiding memory, writing actually facilitates forgetting. As Thamus says to Theuth:

> You provide your students with the appearance of wisdom, not with its reality. Your invention will enable them to hear many things without being properly taught, and they will imagine that they have come to know much while for the most part they will know nothing'.[103]

---

102. Ibid., 270b.
103. Ibid., 275a–b.

This of course is the inspiration for Derrida's famous argument on pharmacology,[104] according to which technics is at the same time a poison and a remedy, further taken up by Bernard Stiegler as the basis for a political programme.[105] Let us highlight here a nuanced difference between Thamus's critique of technics and Zhuangzi's warning. Plato wants to say that by reading along, one may come to know many things, but will not necessarily grasp their truth. For example, one can read a book or watch a video about swimming, but it doesn't mean that one will be able to swim. This is an argument about 'recollection' or 'anamnesis' as the condition of *truth*: writing simply short-circuits this process of anamnesis. Zhuangzi's argument, on the other hand, is rather a straight refusal of any calculation which deviates from *Dao*; through this refusal, Zhuangzi does not seek to affirm reality or truth, but rather to reaffirm freedom.

§10.2. QI AND DAO IN CONFUCIANISM: RESTORING THE LI

In Daoism, then, the unity of *Dao* and *Qi* is exemplified by Pao Ding and his knife. The perfection of the technical tool is also a perfection of living and being, since it is guided by the *Dao*. In Confucianism, though, we find another understanding of *Qi* which seems different from the Daoist one, although they share the same concern for the cosmos and the form of living. In Confucianism, *Qi* often refers to the instruments used in rituals, or *Li* (禮). Indeed, according to the etymologist

---

104. J. Derrida, 'Plato's Pharmacy', in *Dissemination*, tr. B. Johnson (Chicago: University of Chicago Press, 1981): 63–171.

105. See B. Stiegler, *Ce qui fait que la vie vaut la peine d'être vécue: De la pharmacologie* (Paris: Flammarion, 2010) and also *Pharmacologie du Front National* (Paris: Flammarion, 2013).

Duan Yucai (段玉裁, 1735–1815), the right-hand side of of the character for *Li*, that is 豊, is *Qi*; moreover, according to another etymologist, Wang Guo Wei (王國維, 1877–1927), the upper part of 豊 comes from the pictogram for instruments made of jade.[106] During the period of corruption and disruption of morality, Confucius's task was to restore the *Li*. According to a naïve materialist reading of the early twentieth century—naïve in the sense that it is based on a simple opposition between the spiritual and the material—this was to amount to a restoration of feudalism. Precisely for this reason, during the Cultural Revolution the Chinese Marxists attacked Confucianism as a regression and an obstacle to communism.

Li (along with *ren*, 'benevolence') is one of the key concepts in Confucius's teaching. The concept of *Li* is twofold: firstly there is a formal sense in which *Li* defines both the power hierarchy indicated by the artificial objects, *Li Qi* (禮器), and the number of sacrifices performed during the rites. During the Zhou dynasty, *Li Qi* referred to different *Qi* with different functions: cooking utensils, objects made of jade, musical instruments, wine utensils, water utensils, etc. The *Qi* made of jade and bronze were indications of identity and rank in the social hierarchy, including the king and the noble class.[107] But *Li Qi* also refers to a spirit or 'content' that cannot be separated from this formal aspect. This content, for Confucius, is a kind of cultivation and practice that nurtures moral sensibility. In the

---

106. Liu Xin Lan (劉昕嵐), 'On the Origin of Li (論「禮」的起源)', 止善 8 (June 2010), 141–61: 143–4.

107. See Wu Shizhou (吳十洲), *A Study on the Institution of Ritual Vessels during the Zhou Dynasty* (兩周禮器制度研究) (Taipei: Wunan Books, 2003), 417–19. Wu shows that, according to archaeological discoveries, after the Zhou dynasty there was a change from the use of *Li Qi* as funerary objects to *Ming Qi* (明器), meaning that jade and bronze objects were replaced with porcelain substitutes, which also suggests the decline of the *Zhou Li*.

chapter 'Qu Li (曲禮)' of *Li Ji* (禮記, *Book of Rites*), Confucius says that 'the course (of duty), virtue, benevolence, and righteousness cannot be fully carried out without the rules of propriety; nor are training and oral lessons for the rectification of manners complete'.[108] We can understand from this that the moral—that is, one's relation to the heaven—can only be maintained through the practice of *Li*.

Li Ji contains a chapter on *Qi* ('Li Qi [禮器]') which states that 'what is *Li* is convenient to heaven, to the earth, to the gods and the ghosts, as well as the humans; it manages the ten thousand beings (禮也者，合於天時，設於地財，順於鬼神，合於人心，理萬物者也。)'. We might say that, for Confucianism, *Qi* functions to stabilise and restore the moral cosmology through ritual—something we can glimpse in the following example in the chapter '*Li Yun*' ('The Fortune of *Li*'):

> Thus it is that the dark-coloured liquor is in the apartment (where the representative of the dead is entertained); that the vessel of must is near its (entrance) door; that the reddish liquor is in the hall; and the clear, in the (court) below. The victims (also) are displayed, and the tripods and stands are prepared. The lutes and citherns are put in their places, with the flutes, sonorous stones, bells, and drums. The prayers (of the principal in the sacrifice to the spirits) and the benedictions (of the representatives of the departed) are carefully framed. The object of all the ceremonies is to bring down the spirits from above, even their ancestors; serving (also) to rectify the relations between ruler and ministers; to maintain the generous feeling between father and son, and the harmony between elder and younger brother; to adjust the

---

108. '道德仁義，非禮不成，教訓正俗，非禮不備'. *Book of Rites*, tr. J. Legge, <http://ctext.org/liji/qu-Li-i>.

relations between high and low; and to give their proper places to husband and wife. The whole may be said to secure the blessing of Heaven.[109]

As argued by the philosopher Li Zehou (1930–) among others, it is also possible to trace this ritual back to the Xia Shang Zhou dynasties (2070–771 BC) and the shamanic rites associated with them. During the Zhou dynasty, the Emperor formalised the shamanic rites into *Li*, hence they are known as *Zhou Li*. Confucius sought to restore these *Zhou Li* as a resistance against political and social corruption.[110] Thus Li Zehou proposed that *Zhou Li* was 'spiritualised' by Confucius, and then 'philosophised' by Sung and Ming Neo-Confucianism, but that, in this long process, the spirit of the rites—namely, the unification between the Heaven and human beings—was conserved:

> They proceed to their invocations, using in each the appropriate terms. The dark-coloured liquor is employed in (every) sacrifice. The blood with the hair and feathers (of the victim) is presented. The flesh, uncooked, is set forth on the stands. The bones with the flesh on them are sodden; and rush mats and coarse cloth are placed underneath and over the vases and cups. The robes of dyed silk are put on. The must and clarified liquor are presented. The flesh, roasted and grilled, is brought forward. The ruler and his wife take alternate parts in presenting these

---

109. 故玄酒在室,醴醆在戶,粢醍在堂,澄酒在下 。陳其犧牲,備其鼎俎, 列其琴瑟管磬鐘鼓,修其祝嘏,以降上神與其先祖,以正君臣,以篤父子,以睦兄弟,以齊上下,夫婦有所。是謂承天之祜。For the English translation see *Li Ji* (bilingual version), tr. J. Legge, <http://ctext.org/liji/Li-yun>.

110. Li Zehou (李澤厚), *A Theory of Historical Ontology* (歷史本體論) (Beijing: SDX Joint Publishing, 2002), 51.

offerings, *all being done to please the souls of the departed, and constituting a union (of the living) with the disembodied and unseen*. These services having been completed, they retire, and cook again all that was insufficiently done. The dogs, pigs, bullocks, and sheep are dismembered. The shorter dishes (round and square), the taller ones of bamboo and wood, and the soup vessels are all filled. There are the prayers which express the filial piety (of the worshipper), and the benediction announcing the favour (of his ancestors). This may be called the greatest omen of prosperity; and in this the ceremony obtains its grand completion.[111]

While Li Zehou is right to point out this link between *Li* and shamanism, however, it is necessary to bear in mind that, along with Confucianism, the Daoism and Moism that emerged during the same period in China also indicated a rationalisation which marked a break from shamanism.[112] It is possible for the formal aspect of *Li* to dominate its content, and Confucius was aware of this problem. To avert this usurpation of content by form, he emphasises that *Li* is a fundamentally moral practice which starts with individual

---

111. *Li Ji*, <http://www.ctext.org/liji/Li-yun> (italics mine). 作其祝號，玄酒以祭，薦其血毛，腥其俎，孰其殽，與其越席，疏布以冪，衣其浣帛，醴醆以貢獻，薦其燔炙，君以夫人交獻，以嘉魂魄，是謂合莫。然後退而合享，體其犬豕牛羊，實其簠簋籩豆鉶羹。祝以孝告，嘏以慈告，是謂大祥。

112. See YuYing-Shih. 'Between the Heavenly and the Human', in Tu Weiming and M. E. Tucker (eds), *Confucian Spirituality* (New York: Herder, 2003), 62–80. Chinese historians often refer to what the German philosopher Karl Jaspers, in his *The Origin and Goal of History* (tr. M. Bullock [London: Routledge, 2011]) called the Axial Age (*Achsenzeit*). Jaspers claims that new ways of thinking appeared in Persia, India, China, and the Greco-Roman world in religion and philosophy during the 8th to 3rd centuries BC; schools such as Daoism, Confucianism, Moism and others belong to such a 'historical rupture' in knowledge and knowledge production

reflection, extending to outer domains such as family and the state, guided by *Dao*. This is the famous doctrine of *neisheng waiwang* (内聖外王, 'inner sageliness–outer kingliness'). It follows a linear trajectory, as indicated in the Confucian classic *Da Xue* (大學, 'Great Learning' or 'University'): 'investigation of things (格物)', 'extension of knowledge (致知)', 'sincere in thoughts (誠意)', 'rectify the heart (正心)', 'cultivate the persons (修身)', 'regulate the families (齊家)', 'govern well the States (治國)', and 'world peace (平天下)'. In Book XII of *The Analects of Confucius*, we read:

> Yan Yuan asked about *ren*.
>
> The Master said: To master the self and return to *Li* is to be *ren*. For one day master the self and return to *Li*, and the whole world will become *ren*. Being humane proceeds from you yourself. How could it proceed from others?
>
> Yan Yuan said: May I ask how to go about this?
>
> The Master said: If it is contrary to *Li*, don't look at it. If it is contrary to *Li*, don't listen to it. If it is contrary to *Li*, don't utter it. If it is contrary to *Li*, don't do it.[113]

Li is therefore both a set of constraints and a practice that ensures the order of things, so that the perfection of the individual will lead to the perfection of the state. *Dao* is immanent, but one can only know it through self-reflection and through the practice of *Li*. (In the *Analects*, during a dialogue between Confucius and the Prince of Wei Ling, the latter asks about the art of war. Confucius replies that he knows only about *Li*, and nothing about war; and leaves the following day.) But what is this order that *Li* seeks to ensure? A simplistic reading

---

113. Confucius, *The Analects*, 80.

might claim that it is an order socially constructed in favour of the governing class. This is not entirely incorrect, since Confucius emphasises that *Qi* and *Ming* (名, 'name') have to be properly placed so as to maintain order. In *Zuo Zhuan* (400 BC), it is said that a commandant, Yi Xu, rescued the king of the Wei country Sun Huanzi during the war, in order to avoid his being arrested. Sun wanted to give Yi Xu cities as a token of his gratitude. Yi Xu refused, but requested 'to be allowed to be received like a state prince at court, with musical instruments, and to be dressed with the saddle-girth and bridle-trappings of a prince'.[114] Confucius lamented the granting of this request, saying 'Alas! It would have been better to give him many cities. It is only peculiar articles of use, and names, which cannot be granted to another [than those to whom they belong];—to these a ruler has particularly to attend'.[115] As Confucius explained, this is not purely a matter of formality: his reasoning is that *Qi* and *Ming* ensure that those who bear *Ming* and *Qi* should behave properly:

> It is by [the right use of] names that he secures the confidence [of the people]; it is by that confidence that he preserves the articles [*Qi*]; it is in those articles that the ceremonial distinctions of rank are hidden; those ceremonial distinctions are essential to the practice of righteousness; it is righteousness which contributes to the advantage [of the State]; and it is that advantage which secures the quiet of the people. Attention to these things is the condition of [good] government.[116]

---

114.   Zuozhuan (左傳•成公二年), <http://www2.iath.virginia.edu:8080/exist/cocoon/xwomen/texts/chunqiu/d2.14/1/0/bilingual> [translation modified].

115.   Ibid., '年惜也，不如多與之邑，唯器與名，不可以假人，君之所司也'.

116.   Ibid., '名以出信，信以守器，器以藏禮，禮以行義，義以生利，利以平民，政之大節也'.

To summarise, in Confucianism *Qi* has its use in a formal setting, but such use serves only for the purpose of preserving the moral, the heavenly order, and for cultivating great personality; in Daoism, on the other hand, *Qi* plays no such instrumental role, since it is possible to reach the *Dao* by being natural or *zi ran*. *Dao* is metaphysical since it is formless; and in this sense, the metaphysical is the non-technical, the non-geometrical. Even though there are formalised orders in Confucianism, they exist for the purpose of maintaining this superior, formless (or 'above form') *Dao*. The formless is *tian* (the Heaven) and *zi ran*, and it is the formless that has the highest degree of freedom. We might say that the Confucians and the Daoists have different ways of pursuing *Dao*, and that therefore they do not stand in contradiction to one other, but rather complement one other. Mou Zongsan suggests that we characterise Daoism as 'practical ontology' and Confucianism as 'moral metaphysics',[117] in the sense that Confucianism asks the 'what' questions (what is the sage [聖], wisdom [智], benevolence [仁] and rightfulness [義]?), while Daoism asks how to achieve them.[118] For the Daoists, the refusal of mechanical reasoning is a refusal of a calculative form of thinking, in order to stay within the freedom of the inner spirit. We might say that they refuse all efficiency in order to prepare for an opening—a reading of the *Zhuangzi* that resonates superficially with what the late Heidegger calls *Gelassenheit*, which may explain why the Heideggerian critique of technology has found such great resonance among

---

117. Mou Zongsan, *Nineteen Lectures on Chinese Philosophy*, 74.

118. Ibid., 106; Jullien also points out that, for example, 'non-action' is not only a Daoist principle, but is commonly shared in the intellectual tradition: see Jullien, *Procès ou Création*, 41.

Chinese scholars ever since Heidegger affirmed *Gelassenheit* as a possible exodus from modern technology.

This is the ambivalence of *Dao*, then: on one hand, it stands for the completion of technics in the name of nature; on the other, it is also understood as a resistance of the spirit against technics, which always have the potential to contaminate it. Here Heidegger's concept of the truth as *a-letheia* or *Un-verborgenheit* as an access to the open may seem very close to *Dao*; yet, as we shall see below, they are fundamentally different. And indeed, this fundamental difference is one of the reasons it is necessary to conceive of different histories of technics.

§10.3. REMARKS ON STOIC AND DAOIST COSMOTECHNICS

So far, we have tried to trace out a cosmotechnics in Chinese thinking; we have compared it with the Greek concept of *technē*, partly through the writings of Heidegger. It would be too provocative to say that Heidegger was essentially searching for a cosmotechnics, but it is undeniable that the question of *physis* and Being concerns a profound relation between the human and the cosmos. The cosmotechnics that I have sketched above in the traditions of Confucianism and Daoism may seem to some readers similar to Hellenistic philosophy after Aristotle; in particular, the Greco-Roman Stoics' teaching on living in accordance with nature has a clear affinity with the Daoist aspiration to nature (as noted above, Heidegger remained silent on the Stoics, even though the Stoic cosmology seems closer to the Ionian than to the Aristotelian one).[119] There are certainly differences, which

---

119.   Kahn, *Anaximander and the Origins of Greek Cosmology*, 203. Kahn further points out (210) that in the second half of the fourth century, Aristotle rejected cosmogony as a valid explanatory scientific discipline.

we will try to briefly elucidate here. But rather than merely listing these differences, I would like to restate the concept of cosmotechnics developed in the introduction, which hinges on the relation between the cosmos and the moral, as mediated by technics, and to show how we can identify such a cosmotechnics in Stoicism.

A closer reading of the Stoics allows us to see the role played in their thought by *rationality*, which was very much depreciated in Daoism. Both Stoic cosmotechnics and Daoist cosmotechnics propose that we live in accordance with a 'nature'—respectively, *physis* and *zi ran*—and insist that technical objects are only the means towards a more superior end: for Stoics *eudaimonia*, for Daoists *xiao yao* (逍遙, 'free and easy'), and for Confucians *tan dang* (坦蕩, 'magnanimity'). In the first chapter of the *Zhuangzi*, entitled 'Free and Easy Wandering', Zhuangzi describes what he means by freedom with the help of an illustration drawn from Daoist philosopher Liezi:

> If he had only mounted on the truth of Heaven and Earth, ridden the changes of the six breaths, and thus wandered through the boundless, then what would he have had to depend on? Therefore I say, the Perfect Man has no self; the Holy Man has no merit; the Sage has no fame. [120]

Only in following nature rather than attaching oneself to something upon which one will become increasingly dependent can one be free. In *The Analects* (book 7.35), Confucius tells us:

---

120. Zhuangzi, *The Complete Works of Zhuangzi*, 3, '若夫乘天地之正, 而御六氣之辯, 以遊無窮者, 彼且惡乎待哉! 故曰: 至人無己, 神人無功, 聖人無名。'

> The gentleman (*jun zi*) is composed, at peace with things [*tan dang dang*, without worry, unbothered by conflicts and contradictions]. The petty man is constantly fretting, fretting[121]

To be a *jun zi* is to know the will of the Heaven, as the Master says (book 2.4):

> At fifteen I set my mind on learning; by thirty I had found my footing; at forty I was free of perplexities; by fifty I understood the will of Heaven; by sixty I learned to give ear to others; by seventy I could follow my heart's desires without overstepping the line.[122]

Before one can know the will of Heaven, one must study, and only after having attended to one's learning does one become open and free.

In *The Art of Living*, John Sellars draws a contrast between Aristotle's and the Stoics's appropriation of Socrates. According to Sellars, Aristotle attempts to emphasize the relation between philosophy and *logos* in his interpretation of Plato. In the first book of the *Metaphysics*, Aristotle presents Socrates as someone who turns away from nature to ethics, which concerns universals and definitions.[123] Sellars argues that, in doing so, Aristotle underplays the role of philosophy as *askēsis* in Socrates's life and teaching, something that was an inspiration to the Stoic Zeno. The reason, as Sellars points out, is that Aristotle's own philosophical interest was more

---

121. Confucius, *The Analects of Confucius*, 52.

122. Ibid., 20.

123. J. Sellars, *The Art of Living: The Stoics on the Nature and Function of Philosophy* (Bristol: Bristol Classical Press, 2003), 34.

in *logos*.[124] Yet in fact, when Socrates replies to Callicles's question concerning the meaning of 'governing oneself' in the *Gorgias*,[125] he says that it means *sōphrona onta kaì enkratē autòn heautou*, 'mastering one's own pleasures and appetites'.[126] In *Alcibiades I*, Socrates says that the first step in taking care of oneself is to follow the famous Delphic inscription, namely 'know thyself' (*gnôthi seauton*):[127] to take care of oneself is to take care of one's soul in the same way that gymnastics takes care of the body. In the *Apology*, in defence of the accusation against him, Socrates responds:

> You are an Athenian, a citizen of the greatest city with the great-
> est reputation for both wisdom and power; are you not ashamed
> of your eagerness to possess as much wealth, reputation and
> honour as possible, while you do not care for nor give thought
> to wisdom or truth, or the best possible state of your soul.[128]

If the pseudo-arts aim at pleasure, then the genuine arts aim at what is best for the soul[129]—something which could not be demonstrated better than in the picture described in the *Symposium*, in which Socrates sleeps with his arms around the young and beautiful Alcibiades without displaying any sign of sexual arousal.[130]

---

124. Ibid.

125. Plato, 'Gorgias', in *Complete Works*, 491d11.

126. Ibid. See A.A. Long, *From Epicurus to Epictetus* (Oxford: Oxford University Press, 2006), 8.

127. Sellars, *The Art of Living*, 38.

128. Plato, 'Apology', in *Complete Works*, 29e; also cited by Foucault, 'Technologies of the Self', 20.

129. Sellars, *The Art of Living*, 41.

130. Long, *From Epicurus to Epictetus*, 9.

Aristotle's interest in logos and contemplation gives us a different definition of *eudaimonia* than that of the Stoics. In his *Rhetoric*, Aristotle defines happiness as 'prosperity combined with virtue', which consists of internal good (the good of the soul and of the body) and external good (good birth, friends, money, and honour).[131] In the *Nicomachean Ethics* (Book 1, chapter 3), Aristotle describes *eudaimonia* as the *telos* of political science; in the same passage, *eudaimonia*, which is conventionally translated as 'happiness', is identified with 'living well and doing well'.[132] For Aristotle, happiness is related to virtue, yet cannot be guaranteed by virtue. In Book 1 Chapter 7, he explains what he means by good: good is defined by the final end internal to the action itself—for example, in medicine it is health, in strategy, victory, in architecture, a house. Aristotle concludes that 'if there is an end of all that we do, this will be the good achievable by action, and if there are more than one, these will be the goods achievable by action'.[133] Virtue is not the guarantee of happiness, since man, unlike vegetables and animals, is endowed with rational principles. Rationality is what exceeds mere functionality and aims at the most desirable good. The human good, says Aristotle, 'turns out to be activity of soul in accordance with virtue, and if there were more than one virtue, in accordance with the best and most complete'.[134] Thomas Nagel suggests that this move is an affirmation of reason above other functions such as perception,

---

131. Aristotle, 'Rhetoric', in *Basic Works*, 1360b26-28.

132. Aristotle, 'Nicomachean Ethics', in *Basic Works*, 1095a19.

133. Ibid., 1097a23-25.

134. Ibid., 1098a16-18.

locomotion, and desire, which support reason, while reason is
not subordinated to them.[135]

The relation between Aristotle and the Stoics is still a
subject of debate. A.A. Long has shown that Aristotle's conception of *eudaimonia* had a direct influence on the Stoics, and
David E. Hahm has demonstrated that the Stoics' cosmologies
were influenced more by Aristotle than by Plato's *Timaeus*. A
key difference that is agreed upon among classical scholars,
however, is that, unlike Aristotle, for whom the external good
plays a role in the realisation of *eudaimonia*, for the Stoics,
eudaimonia consists entirely in ethical virtue: good or bad,
pleasure or its absence, are matters of indifference.[136] And
here lies the Stoics' most important axiom: as defined by Zeno,
it consists in 'living in agreement'; which Cleanthes amends
to 'living in agreement with nature', and Chrysippus, 'living in
accordance with experience of what comes about by nature'.[137]
Julia Annas refers to this nature as 'cosmic nature':[138] once
more, virtue has its perfect model in the organization of the
universe, and the human being is part of the cosmic nature,
and therefore the cosmos is the perfect model for virtue, in
a way that appears similar to the Chinese thinking of *Dao*.

135.   T. Nagel, 'Aristotle on Eudaimonia', in A. Rorty (ed.), *Essays on Aristotle's Ethics* (California: University of California Press, 1980), 11.

136.   A.A. Long, 'Stoic Eudaimonism', in *Stoic Studies* (Berkeley, CA: University of California Press, 2001), 182.

137.   J. Annas, *The Morality of Happiness* (Oxford: Oxford University Press, 1995), 168; cited from Arius Didymus, in Stobaeus, *Eclogae (Selections)* Book II. 85.12–18. The citation continues with Diogenes and Archedemus, 'being reasonable in the selection and counter-selection of the things according to nature'; Archedemus: 'living so as to make all of one's due actions complete'; and Antipater: 'living selecting things according to nature and counter-selecting things contrary to nature invariably'. To which he also added: 'doing everything one can invariably and unalterably towards obtaining the things that are preferable according to nature'.

138.   Annas, *The Morality of Happiness*, 159.

But how do the Stoics pass from *physis* to the moral? The Stoic cosmos is one limited, spherical body surrounded by the infinite void. A common reading has it that they follow the Heraclitean model of the cosmos, which sees it as being generated by the mixing of material with fire, which is breath and vital heat. The cosmos repeats itself in an identical cycle, in which fire is transformed into other elements and then returns to itself. There is a logic to be found in the cosmos that is produced by Reason, and Reason 'cannot produce one which is either better or worse'.[139] In Cicero's *On the Nature of the Gods* we find a precise description of the passage from the physical to the moral, in which Reason becomes divine:

> And contemplating the heavenly bodies the mind arrives at a knowledge of the gods, from which arises piety, with its comrades justice and the rest of the virtues, the sources of a life of happiness that vies with and resembles the divine existence and leaves us inferior to the celestial beings in nothing else save immortality, which is immaterial for happiness.[140]

---

139. P. Hadot, *What is Ancient Philosophy?* tr. M. Chase (Cambridge, MA: Harvard University Press, 2004), 130; however, we also note that this Heraclitean reading of Stoic cosmology is contested by some authors such as Hahm, who argues for the influence of Plato, and even more so of Aristotle, since the Stoics took up Aristotle's classification of five elements in *On the Heavens*—fire, air, water, earth and ether (the celestial element)—integrating them into a biological model of the cosmos as a living being. See Hahm, *The Origins of Stoic Cosmology*, 96–103. It is also said that, for Zeno, the constitutive element of the cosmos is fire, for Cleanthes, heat, and for Chrissypus, *pneuma*: see J. Sellars, 'The Point of View of the Cosmos: Deleuze, Romanticism, Stoicism', *Pli* 8 (1990), 1–24: 15n70.

140. Cicero, 'De Natura Deorum', in *Cicero in Twenty-Eight Volumes*, vol. XIX, tr. H. Rackham (London: William Heinemann, 1967), II, LXI, 153; also quoted by Goldschmidt, *Le système stoïcien et l'idée de temps*, 67.

The mediation between the two realms consists of the core idea of what the Stoics call *oikeiosis*. The Stoic morality is not a categorical moral obligation, although it involves self-reflection and self-restriction; to live in agreement with nature requires both contemplation *and* interpretation. Interpretation means firstly to place oneself in relation with beings through contemplation, and secondly to give value to them. These values are not arbitrary, as Émile Bréhier pointed out: 'value is not what gives measure, but what is to be measured; what gives the measure is being itself [...] in other words: axiology supposes ontology and doesn't replace it'.[141]

Gábor Betegh has proposed that the Stoics, particularly Chrysippus, had convincingly integrated cosmic nature into the foundation of their ethical theory. Betegh's position opposes Julia Annas's argument in *The Morality of Happiness* that the Stoics' ethical theory was developed 'prior to' and 'independently of' their physical and theological doctrines. If this is true, then, since physics would be a mere supplement for deepening our understanding of ethics, we would be mistaken in setting out from cosmic nature in order to understand the nature of Stoic ethics.[142] We have already encountered a similar argument in our discussion of Mou Zongsan's critique of Dong Zhongshu's cosmocentrism; however, as we pointed out, morality is not possible without taking the external environment into

141. E. Bréhier, 'Sur une théorie de la valeur dans la philosophie antique', *Actes du IIIe Congrès des Sociétés de Philosophie de langue française* (Louvain: Editions E. Nauwelaerts, 1947), cited by Goldschmidt, *Le système stoïcien et l'idée de temps*, 70.

142. G. Betegh, 'Cosmological Ethics in the *Timaeus* and early Stoicism', *Oxford Studies in Ancient Philosophy* 24 (2003): 273–302: 275; Annas, *The Morality of Happiness*, 166.

account, since it is being-in-the-world that is the condition of ethical thought.

Betegh showed that Plato's *Timaeus* has an important influence on Chrysippus's theory of *telos*. The long passage in the *Timaeus* upon which Betegh develops his thesis reads as follows:

> Hence if someone has devoted all his interest and energy to his appetites or to competition, all his beliefs must necessarily be mortal ones, and altogether, so far as it is possible to become *par excellence* mortal, he will not fall the least bit short of this, because it is the mortal part of himself that he has developed. But if someone has committed himself entirely to learning and to true wisdom, and it is these among the things at his disposal that he has most practised, he must necessarily have immortal and divine wisdom, provided that he gets a grasp on truth. And so far as it is possible for human nature to have a share in immortality, he will not in any degree lack this. And because he always takes care of that which is divine, and has the *daimon* that lives with him well ordered [εὖ κεκοσμημένον τὸν δαίμονα], he will be supremely happy [εὐδαίμονα]. Now for everybody there is one way to care for every part, and that is to grant to each part its own proper nourishments and motions. For the divine element in us, the motions which are akin to it are the thoughts and revolutions of the whole world. Everyone should take a lead from these. We should correct the corrupted revolutions in our head concerned with becoming by learning the harmonies and revolutions of the whole world, and so make the thinking subject resemble the object of its thought, in accordance with its ancient nature; and by creating this resemblance, bring to

fulfillment [τέλος] the best life offered by gods to mankind for
present and future time.[143]

Here, in an apparent echo of the relation in Chinese thought
between the human and the Heaven, we find a parallel
between the structure and organisation of the individual soul
and the world soul[144]—a kind of 'analogy'. Yet in Plato the relation is not truly analogical, since the human being is also within
nature and is a part of the whole. It is possible to bring the
rational part of the soul into order and harmony when the soul
internalizes the cosmic harmony. This process is initiated with
oikeiōsis, usually translated as 'appropriation'. A reconstruction
based on Cicero's *De Finibus Bonorum et Malorum*, Seneca's
Epistulae morales ad Lucilium, and Diogenes Laertius's report,
gives us the general picture:[145] the Stoics believe that man and
animal are both given the ability to distinguish what is proper
(*oikeion*) to their constitution (*sustasis*) and preservation,
from what is not proper or strange (*allotrion*): according
to Diogenes Laertius, Chrysippus remarks that it would not
be reasonable if, having created an animal, nature gave it
no means of self-preservation. However, a second stage is
necessary in which such *oikeiōsis* demands insight through
which one's action can be guided by reason. The perfection
of reason is identified with nature, since nature prescribes
virtuous behaviours.

---

143. Plato, *Timaeus*, cited in Betegh, 'Cosmological Ethics', 279.

144. Betegh, 'Cosmological Ethics', 279.

145. The following description is given by G. Striker, 'The Role of Oikeiōsis in
Stoic Ethics', in *Essays on Hellenistic Epistemology and Ethics* (Cambridge:
Cambridge University Press, 1996), 281–97: 286–7.

The Stoics' 'art of living' is, as the word 'art' would suggest, a *technē*. Annas has suggested that 'Stoics regard virtue as a kind of skill (*technē*), and that skill is an intellectual grasp that builds up, becoming ever firmer, through trial and error. As they put it, virtue is the skill concerned with a life productive of happiness'.[146] We might look here at Zeno's formal definition of happiness as a 'good flow of life';[147] and *technē* as 'a system of apprehensions unified by practice for some goal useful in life'.[148] These definitions are of course not straightforward; however they do suggest that the *technē* that aims for virtue facilitates the proper flow of life,[149] including dealing with anger, mercy, revenge, etc. Marcus Aurelius, for example, advises us to contemplate an object, to imagine that it is dissolving and transforming, rotting and wasting away. Hadot points out that this exercise of imagining universal metamorphosis is linked to the meditation on death, such that it 'leads the philosopher to give loving consent to the events which have been willed by that Reason which is immanent to the cosmos.'[150]

---

146. Annas, *Morality of Happiness*, 169.

147. Long, 'Stoic Eudaimonism', 189.

148. Sellars, *The Art of Living*, 69; One should also pay attention to the term 'system' (*systema*), as F.E. Sparshott has pointed out: for the Stoics, it is necessarily something coming 'out of' (*ek*) the actuality (and in this sense, is unlike the Platonic concept of the Idea); *technē*, for the Stoics, is a system constituted from grasping (*ek katalēpseōn*). See F.E. Sparshott, 'Zeno on Art: Anatomy of a Definition', in J.M. Rist (ed.), *The Stoics* (Berkeley, CA: University of California Press, 1978), 273–90.

149. Sellars suggests three types of technics here: (1) a productive technics, which has an end product; (2) a performative technics, whose product is less important than the act itself; (3) a stochastic technics, which aims for the best, but is not necessarily guaranteed, e.g. medicine. See *The Art of Living*, 69–70.

150. Hadot, *What is Ancient Philosophy?*, 136.

With all of this in mind, we might list the following differences between Daoism and Stoicism in terms of 'living in agreement with nature':

Cosmology: the Stoics model the cosmos as organism (and one might speak here of a cosmobiology or cosmophysiology),[151] something that is not evident in Daoism, where there is an organic organization of the universe, but where it is not presented as an animal, but is instead guided by *Dao*, which is modelled on *zi ran*;[152]

Divinisation: for the Stoics, the cosmos is related to the divine qua lawgiver, while this role of the lawgiver or creator is not found in ancient Chinese thinking;

Eudaimonia: the Stoics value rationality highly since it is what leads to *eudaimonia*, and the human plays a specific role in the universe owing to its rationality; Daoists may recognize the former, but reject the latter, since Dao is in all being, and freedom can only be achieved through *wu wei* (non-action);

Rationality: for the Stoics, to live with nature is to *develop* rationality; for Daoists, it is rather a matter of restoring one's original spontaneous aptitude.[153]

---

151.   See Hahm, *The Origins of Stoic Cosmology*, Chapter 5, 'Cosmobiology', 136–84.

152.   E.T. Ch'ien argued that, although the Stoic *hēgemonikon* or the Platonic world soul seems to be based on a coordinative logic, it is in fact still a subordinative logic: see Ch'ien, *Lectures on the History of Chinese Thought* (中國思想史講義), 220.

153.   The last three points are derived from Yu Jiyuan, 'Living with Nature: Stoicism and Daoism', *History of Philosophy Quarterly* 25:1 (2008), 1–19.

The above remarks aim to show that the relation between the cosmos and the moral in the Stoicism and Daoism are mediated by different technics, which in turn belong to what I term cosmotechnics. These relations are established in different ways, and in fact define different modes of life. In 'Technology of the Self', Foucault gives various examples of the Stoics' practices: letters to friends and disclosures of the self (Marcus Aurelius, Seneca, etc.); examination of the self and conscience; the *askēsis* of remembering truth (not discovering truth).[154] The Greeks classified techniques in two main forms: *meletē* and *gymnasia*. *Meletē* means meditation, in which one uses imagination to help oneself to cope with a situation, e.g. imagining the worst scenarios, perceiving that undesirable things are already taking place, refusing the conventional perception of suffering (e.g. illness). *Gymnasia*, on the contrary, consists of bodily exercise, such as strenuous sporting activities.[155] We may want to ask: How do these exercises have their ground in the understanding of virtue revealed by cosmic nature? This is not the point that concerns Foucault, who is interested in the history of self-disclosure, but it is precisely the question that needs to be addressed in our inquiry into cosmotechnics.

The integration of the Stoics' practice into the early Christian doctrine, as Foucault pointed out, brought about a profound transformation. If 'knowing yourself' was a consequence of 'taking care of yourself' in Stoicism, in Christian doctrine it became directly linked to the disclosure of the self as sinner and penitent.[156] Foucault listed two main techniques,

---

154. Foucault, 'Technologies of the Self', 34.

155. Ibid., 36–7.

156. Ibid., 41.

exomologēsis, which operates by showing shame and humility, exhibiting modesty, not as a private practice, as in Seneca, but through *publicatio sui*; and *exagoreusis*, which is based on two principles, obedience and contemplation, so that self-examination leads to the recognition of God. A transformation in Daoist practices was also observed when Daoist teaching (*dao jia*) became adopted into a religion (*dao jiao*): meditation, the martial arts, sexual practices, alchemy, etc.; but unlike what took place with the selection and transformation of Hellenistic teaching in Christian doctrine, in Daoism the essence of the thought of Laozi and Zhuangzi remain intact; and Daoism also effectively absorbed the Confucian understanding of the 'resonance between the heavens and the human' into its teaching.

This should contribute toward a comprehensive understanding of the concept of cosmotechnics and the necessity to open up the concept of technics and the history of technics. In the rest of this part, I will sketch the transformation of the Dao-Qi relations in China, before we come to Part II to access its significance for the understanding of modernity and modernization.

§11. QI-DAO AS RESISTANCE: THE GU WEN MOVEMENT IN THE TANG PERIOD

It has been suggested that we can systematically understand Chinese philosophy through the analysis of the dynamics between *Qi* and *Dao*. And the attempt to reaffirm the unification of *Qi* and *Dao* has been omnipresent in every epoch, particularly at moments of crisis. According to historians Jin Guantao and Liu Chingfeng, the Wei Jin dynasty (220–420 CE) is one of the two most interesting periods for research on the history of Chinese thought, since it was the period when Buddhism came to China, provoking an internal transformation which finally led to a unification of Confucianism, Daoism, and Buddhism, and therefore may be said to have shaped what remained the dominant tradition of philosophy in China up until the mid-nineteenth century. The other period is that which followed the 1840s, meaning the period of modernisation in China, which we discuss in detail below. We will see that, during both of these periods, the unification of *Qi* and *Dao* was reaffirmed as a resistance against external threats (namely Buddhism and Western culture), but that different historical contexts yielded different dynamics between *Qi* and *Dao*. We should also add another period prior to these two: the decline of the Zhou dynasty (1046–256 BC). As Mou Zongsan suggests, the emergence of Confucianism and Daoism was a response to the corruption of the system of *Li* (rites) and *Yue* (music) established by the King Wen of Zhou (1152–1056 BC), which led to the corruption of the moral.[157] This transformation of the *Qi-Dao* relation is crucial to understanding the question of technology in China.

During the Tang dynasty (618–709 CE), Buddhism became the dominant religion in China and the official religion or belief of the government. During the mid-Tang period, the Confucian movement was reinitiated as a resistance against Buddhism—which, to the eyes of intellectuals such as Han Yu (韓愈, 768–824) and Liu Zongyuan (柳宗元, 773–819), was mere superstition. The Tang was the most prosperous period in Chinese history, and probably also one of the most open, during which exchanges between China and neighbouring countries, including royal marriage, were allowed. The anti-Buddhist

157. See Liu Shuhsien (劉述先, 1934–2016), *On Contemporary Chinese Philosophy* (當代中國哲學論) vol. 1 (Hong Kong: Global Publishing Co. Inc., 1996), 192.

movement consisted of two parts: resistance against the superstitions brought in by Buddhism and Daoism as religions; and an effort to reestablish the Confucian teaching of *Dao* (道學) which had been lost since the Han dynasties. This was known as the Gu Wen movement (古文運動), where *Gu* means ancient, and *Wen* means writing. It proposed that writing should enlighten *Dao*, rather than focusing on style and form. During the Wei Jin period, pianwen (駢文, literally 'parallel writing'), characterised by a flamboyance of vocabulary and the parallel form of sentences, was the dominant style of writing. According to Han and Liu, the leaders of the Gu Wen movement, pianwen had deviated from *Dao* in the sense that it had become a superficial aesthetic enterprise. The Gu Wen movement was an attempt to reestablish the ancient style of writing, but also the ancient Confucian teaching. It took as its slogan 'writing enlightens *Dao* [文以明道]'—meaning that writing takes on the role of a specific form of *Qi* capable of reestablishing the unity between *Qi* and *Dao*.

The thrust of this movement can be seen retrospectively as an attempt to reestablish Confucianism as the centre of Chinese culture. But what does centre, or *Zhong* (中), mean here? *Zhong* has a double meaning, one that helps us to distinguish between Han Yu and Liu Zongyuan; more importantly, this double meaning shows that a 'pure', 'original' Confucian teaching cannot be recovered, since *Dao* is not a static, eternal being, and was also influenced by Buddhism. On the one hand, there is the Confucian classic *Zhong Yong* (中庸, 'Doctrine of the Mean'), which emphasises the value of *Zhong*, meaning not to lean to any extreme, to act properly; on the other hand, there is also *Zhong Guan* (中觀), a concept developed by Nāgārjuna, which sees the *Kong* (空, 'void') as the permanent and authentic form of existence, and other phenomena as only illusions, mere phenomena.[158] Han Yu leans more towards the first meaning of *Zhong*, Liu Zongyuan towards the second, since he was more sympathetic to Buddhism. As Han Yu explains his concept of the *Dao* in his article '*Yuan Dao*' (原 道, 'Essentials of the *Dao*' or, more literally, 'Origin of the *Dao*'),

> What were the teachings of our ancient kings? To love univer-
> sally, which is called humanity; to apply this in the proper manner,
> which is called righteousness; to act according to these, which
> is called the Way [*Dao*]; to (follow the Way and) become self-
> sufficient without seeking anything outside, which is called virtue.
> The *Book of Poetry*, the *Book of History*, the *Book of Changes*
> and the *Spring and Autumn Annals* are their writings; rites
> and music, punishments and government, their methods. Their
> people were the four classes of scholar-officials, farmers, arti-
> sans, and merchants; their relationships were those of sovereign
> and subject, father and son, teacher and friend, guest and host,
> elder and younger brother, and husband and wife. Their clothing
> was hemp and silk; their dwelling halls and houses; their food
> grain and rice, fruit and vegetables, fish and meat. Their ways
> were easy to understand; their teachings simple to practice.[159]

Han Yu's interpretation of *Dao* came to be considered conservative and regressive by the reformers of the late Qing

---

158. One can understand the reason why *Zhong* (central) is also *Kong* (void) from the eight non-central forms: no birth no death, no continuity no discontinuity, no unity no difference, no incoming and no outgoing (不生也 不滅, 不常亦不斷, 不一亦不異, 不來亦不出), see Jing Guan-tao and Liu Ching-feng, *Ten Lectures*, 190.

159. Cited by Chen Joshui, *Liu Tsung-yuan and Intellectual Change in T'ang China 773–819* (Cambridge: Cambridge University Press, 1992), 121.

dynasty (1644–1912), since he wished to reinstate feudal-ism[160]—the same reason why Confucius would later attract the criticism of the communists during the Cultural Revolution. In contrast, the Buddhism of Zhong Guan remains for Liu Zongyuan a guiding principle for the development of a unified cosmological thinking which, against the conception of the unity of the Heaven and the human developed during the Han dynasty, would separate them into the supernatural and the natural, the superstitious and the spiritual.[161] The formation of the world is to be found in the world itself, and no transcendence or first cause need be sought. Here we see a thinking that is very close to—if not an actual precursor of—the Neo-Confucianism of the Sung dynasty.[162] The primary world-constituting element that Liu calls *Yuan qi* (元氣) is both a material and a spiritual being—not so far from Sung Neo-Confucianism's conception of *ch'i* theory (氣論).

However, despite the differences between Han Yu and Liu Zongyuan, the overall significance of their movement was to reconstitute the unity between *Qi* and *Dao*. The unity of *Qi-Dao*, explicitly expressed in the Gu Wen movement as the relation between writing and *Dao*, reaffirms the cosmological and moral

---

160. Wu Wenzhi (吳文治), *Biography of Liu Zongyuan* (柳宗元評傳) (Beijing: Zhonghua Book Company, 1962), 188–9.

161. Liu does not believe in the will of the Heaven, and dismisses the ancient interpretation of the relation between winter and punishment as mere superstition. Thunder may break a rock, he says, and when winter comes, trees and herbs die, but one cannot see these as punishments, since the rock and the tree are not criminals. See Luo Zheng Jun (駱正軍), *New Interpretation of Liu ZongYuan's Thought* (柳宗元思想新探) (Changsha: Hunan University Press, 2007), 95.

162. This point is debatable, however; readers interested in this subject are referred to the work of the historian Chen Joshui on this subject, *Liu Tsung-yuan and Intellectual Change in T'ang China*.

order, as well as the Daoist aspiration to *zi ran*, which is evident in much of Liu's prose. A parallel development during the Tang dynasty may add something relevant to this reaffirmation of the *Qi-Dao* unity in everyday life—namely, what historians Jin Guantao and Liu Chingfeng call the 'reason of common sense' (常識理性). According to Jin and Liu, since Wei Jin there has been a tendency to absorb sophisticated philosophical concepts into everyday practice as if they are common sense. This explains the rapid dissemination of Buddhism in Chinese culture (though a full integration would take a thousand years because of the incompatibility between the systems) and the development of quasi-religious forms of Confucianism and Daoism. One of the powerful examples they give is that of Zen Buddhism, since for Zen Buddhism it is not essential to read and interpret the ancient scripts (indeed, many of the great masters cannot even read). But this also characterises the difference between Chinese Buddhism and Indian Buddhism, since for the former, *Dao* is in everyday life, and therefore everyone is capable of becoming Buddha, whereas this is not necessarily the case for the latter. In other words, there is a certain line of thought that implies that *Dao* is not to be sought anywhere else than in everyday life. This 'reason of common sense' is further developed in Sung Ming Neo-Confucianism.

§12. THE MATERIALIST THEORY OF CH'I IN EARLY NEO-CONFUCIANISM

Up to this point, we have only discussed the usage of *Qi*, but not the production of *Qi*. What is the role of *Qi* in the moral cosmology, or moral cosmogony? Moral cosmology attained new heights in Sung and Ming Confucianism,[163] but a 'materialist

---

163. See Mou Zongsan, *Questions and Development of Sung and Ming Confucianism*.

thinking' also emerged out of this context, in which another element was reintroduced so as to elaborate the cosmogony, namely *ch'i* (氣). A materialist theory of *ch'i* was developed by one of the first Neo-Confucians, Zhang Zai (張載, 1020–1077,), and was integrated into the work of Song Yingxing (宋應星, 1587–1666), author of the encyclopaedia of technologies published in 1637 during the Ming dynasty.

What exactly is this *ch'i*, which may be familiar to readers who have some knowledge of Tai Chi and Chinese medicine? It is not simply material or energetic, but is fundamentally moral. We must recognise that Sung Ming Neo-Confucianism was a continuation of the resistance against Buddhism and superstitious Daoism. It centred on a metaphysical inquiry that sought to develop a cosmogony compatible with the moral, and which emerged from the reading of two classics, namely *The Doctrine of Mean* and *Yi Zhuan* (seven commentaries on *Zhou Yi—The Book of Changes*), which in turn came from the interpretation of the *Analects of Confucius* and *Mencius*.[164] Mou Zongsan suggests that the contribution of Sung and Ming Neo-Confucianism could be understood as 'the penetration of the moral necessity to such an extreme that it attains the highest clarity and perfection'.[165] This consists in the unification of 'ontological cosmology' and morality through the practice of *ren* (仁, 'benevolence') and the full development of *xing* (性, 'inner possibility' or 'human nature').[166]

---

164. Ibid., 99.

165. Mou Zongsan, Collected Works 5, *Moral Creative Reality: Mind and Nature*, Vol I(心性與體性) (Taipei: Linking Publishing, 2003) 120, '把道德性之當然滲透至充其極而已達致具體清澈精誠惻怛之圓而神奇之境' (this sentence is almost untranslatable).

166. Ibid., 121, '在形而上(本體宇宙論)方面與道德方面都是根據踐仁盡性的'.

It is not our intention to fully document the thought of the Sung-Ming Confucians, however, but rather to understand the relation between *Qi* and *Dao* during this particular period of Chinese philosophy. Indeed, the three volumes of Mou Zongsan's *Moral Creative Reality: Mind and Nature* (心體與性體, 1968–9) provide a very systematic and historic exposition of the subject matter, which will not be easily surpassed by any future work. Here we seek only to prepare readers to understand some basic ideas which are essential for our own interpretation. The first thinker of the moral cosmogony in Neo-Confucianism is considered to be Zhou Dunyi (周敦頤, 1017–1073) who developed a model based on a diagram of *Taiji*, in which *Wuji* (無極, 'without pole or chaos') engenders *Taiji*, *Taiji* gives movement, which is *Yang*, *Yang* at its limit becomes rest, and rest produces *Yin*. When *Yin* arrives at its extreme, movement reappears. *Yin* and *Yang* yield *Wu Xing* (五行, 'five phases' or 'five movements'), and the movement of *Wu Xing* engenders the ten thousand beings. Zhou Dunyi suggests that the sage developed benevolence and rightfulness in correspondence to the *Yin* and *Yang*, soft and hard, and that therefore his moral stance is identified with heaven and earth.[167]

Zhang Zai continued in this pursuit of the relation between cosmogony and the moral by further developing the concept of *ch'i*. As we have seen, *ch'i* is the elementary component of the cosmos, and all beings are the actualisation of *ch'i* according to its internal movement, which is called *shen* (神, 'spirit').

---

167. Ibid, 376, Zhou cites the Sho Gua from *I Ching*, quoted above in §7: 'The Dao of the heaven is *Yin* and *Yang*, that of earth is soft and hard, that of humans is benevolence and rightfulness. (故曰：立天之道曰陰與陽，立地之道曰柔與剛，立人之道曰仁與義。又曰：原始反終，故知死生之說。)'.

The dynamic process that envelops this great harmony is *Dao*.[168] Zhang Zai calls this process of individuation *ch'i hua* (氣化, 'transformation of *ch'i*'). We should pay attention here to the word *hua* (化), which does not denote a sudden movement like a quantum leap, which would be called *bian* (變), but rather a slow movement that can be likened to the changing of the shape of a cloud in the sky.[169] In simple terms, what underlies this theory of *ch'i* is a monism which furnishes the foundation for the coherence between *cosmology* and the *moral*. With this monism of *ch'i*, Zhang Zai was able to claim that heaven and earth, sun and moon, other human beings and the ten thousand beings are all connected to the *I*.[170] One therefore has a *moral obligation* towards the ten thousand beings (*wan wu*, 萬物), and in turn the ten thousand beings are part of the 'I' (民吾同胞，物吾與也).[171] Once again we return to the core of the Confucian project, namely a *moral cosmology*.

In parallel to *ch'i*, there were also two other schools in Sung Neo-Confucianism, those of *Li* (理, reason) and *xin* (心, 'heart' or 'mind').[172] However, it seems to me that these schools did not take technics into account, and an understanding of technics in relation to metaphysics only becomes more visible in the thought of Song Yingxing (1587–1666).[173] Indeed, it is

---

168. Chen Lai (陳來), *Sung Ming Li Xue* (宋明理學) (Shen Yan: Liao Ning Education Press, 1995), 61–62.

169. *Bian* is also considered to be *Yang*, and *Hua* is considered to be *Yin*.

170. Ibid., 74, "視天下無一物非我".

171. Zhang Zai, *Zheng Meng* (正蒙), with commentary by Wang Fuzhi (王夫之) (Shanghai: Ancient Works Publishing, 2000), 231.

172. Note that there was also another school called 'shu', meaning 'counting', advocated by an early Sung Neo-Confucian, Shao Yung (邵雍, 1011–1077), but we do not have space to cover it here.

173. It is also true that, in the *Zhuangzi*, *ch'i* already played a significant role; however the relation between *Dao*, *Qi*, and *ch'i* was not clear.

not difficult to notice how the focus on *Li* and *Dao* led to a tendency to separate *Qi* and *Dao*: for example, Zhou Dunyi transformed the credo that 'writing enlightens *Dao* (文以明道)' into 'writing conveys *Dao* (文以載道)'. 'Conveying', of course, implies that the two can be separated, since the *Qi* of writing, in this case, is only a vehicle—that is to say, it is merely functional. The *Xin* school tended to see all changes of the universe to be comprehended in the infinite *Xin*, meaning that it sees the *xin* as the absolute and the ultimate possibility, and hence rarely gives any proper role to technics.

§13. QI-DAO IN SONG YINGXING'S ENCY-CLOPAEDIA DURING THE MING DYNASTY

The achievement of Song Yingxing is very significant, since his is probably the first theory to take the role of *Qi* in the process of the individuation of technical and physical beings to a metaphysical level, where *Qi* finds its proper role. In a supplement to the moral cosmogonies developed by the Neo-Confucians, Song rendered explicit the role of cosmotechnics by situating himself within the thought of Neo-Confucianism. In order to appreciate the importance of Song's work, let us give a brief description of what happened after the Tang dynasty.

The Song dynasty (960–1279) was a period of great technological development, bringing with it, for example, the development of the navigational compass, the development of gunpowder and its military application, and the invention of movable type for printing—named as the 'three great inventions' by Francis Bacon in his 1620 *Instauratio magna*. The Yuan dynasty (1271–1368) or Mongolian Empire that followed reached Europe with its horses and warriors, accelerated exchange between East and West—today we are told that Marco Polo came to China during this period. The epoch of Song Yingxing, namely the Ming dynasty (1368–1644), was the period when the development of science and technology, as well as aesthetics, scaled new heights: the first telescope was constructed, Zheng He and his team sailed to Africa, and Euclid's geometry was translated into Chinese.

Song's work didn't come from nowhere, then, but embodied the spirit of his times. His encyclopaedia *Tian Gong Kai Wu* (天工開物, 'The Exploitation of the Works of Nature'), published in 1636, consists of eighteen sections detailing different techniques, including agriculture, metallurgy, and the manufacture of arms. The detailed entries together with commentaries, come from the author's observations during his travels and other research. *Tian* (天, 'Heaven') is the synonym for the cosmological principles that govern all the changes and the emergence of beings. *Tian Gong Kai Wu* is an attempt to understand these principles, and to describe the way in which human intervention in everyday production is compatible with the principle of Heaven.

Song Yingxin's encyclopaedia arrived almost a hundred years before the *encyclopédie* of Jean d'Alembert and Denis Diderot in France and the encyclopaedia of W. & R. Chambers in England. The historical context is very different, of course. Enlightenment encyclopaedism in Europe presents us with a historically new form of the systematisation and dissemination of knowledge which, unlike the Tian Gong Kai Wu, detached itself from 'nature'. Martine Groult has pointed out that, at this particular moment, history is detached from the life of the king, and philosophy is detached from theology.[174] Philosophy is liberated and becomes dominant, participating in different

---

174. M. Groult, 'L'encyclopédisme dans les mots et les choses: différence entre la cyclopaedia et l'encylopédie', in *L'encyclopédisme au XVIIIe siècle: actes du colloque, Liège, 30-31 octobre 2006*, 170.

disciplines and producing a philosophy of relations (*rapports*).[175] In this context, the liberty of philosophy becomes fundamental to Enlightenment values, as defended, for example, by Kant in *Conflict of the Faculties*, where he shows that philosophy, as the lower faculty in the German academic system in comparison to the three 'higher faculties' of theology, law, and medicine, should have the highest liberty. The context in China was quite different: the author of the *Tian Gong Kai Wu* was not known as a philosopher—he only passed the public examination to become a public servant after several failed attempts, when he was already quite old. Later, he took up a very lowly position in the government, and wrote the encyclopedia while living in poverty. Yet what is similar in the two cases is the decisive role of philosophy in the systematisation of technology. And in both cases, it is philosophical thought which—as well as being a sort of 'meta'-thinking, beyond all disciplines—serves to bring different varieties of knowledge into convergence.

It was not until the 1970s that other writings by Song Yingxing were rediscovered, including several important texts such as *Tan tian* (談天, 'On Heaven') and *Lun qi* (論氣, 'On *ch'i*'). In these texts, the connection between technics and the (by then) dominant metaphysics (i.e. Neo-Confucianism) becomes evident. Sung's metaphysics centred around the work of Zhang Zai, briefly mentioned above. Zhang proposes a monism of *ch'i* to explain both cosmogenesis and moral cosmology. In his posthumous work *Zheng Meng* (正蒙), Zhang writes that '*tai he* (great harmony) is called *Dao* [太和之謂道]'. Zhang proposes that Dao is the process of the movement of *ch'i*, and therefore maintains that 'transformation of *ch'i* is known as *Dao* (由氣化，有道之名)'. He contends that 'all that has form exists, all that exists has a phenomenon, all phenomena are *ch'i*,[176] and further that 'knowing that the void is *ch'i*, then being or nothing, hidden or evident, transformation of *shen* (神, 'spirit')[177] or living, can all be known'.[178] Zhang Zai thereby seeks to show that even the void consists of *ch'i*—that the latter is not necessarily related only to phenomena but can also be invisible.[179] Zhang Zai's theory of *ch'i* became a focus of debate concerning the autonomy of *ch'i*: does *ch'i* in itself already carry the principles of movement, or does it require external principles and motivations to regulate its movement?

Contemporaries of Zhang Zai argued that one should separate *ch'i* and *Dao*, since *Dao* is beyond form and appearance. Therefore one should identify *Dao* with *li* (理, 'reason' or 'principle') rather than with *ch'i*. The Cheng brothers[180] presented a counter-proposal according to which 'what has form is *ch'i*, what is formless is *Dao* (有形總是氣，無形只是道)'. Zhangzai's *ch'i* and the Chengs' *li* were both adopted in the theory of Zhu Xi (朱熹 1130–1200), but here *ch'i* is equated with *Qi* (器), and *li* (理) is what is beyond the form—as he says, 'between heaven and earth, one finds *li* and *ch'i*. *Li* is that which is above form, the ground of living beings. But *ch'i* (氣) is below the form, that is *Qi*, possessed by living beings'.[181]

---

176. '凡可狀皆有也，凡有皆象也，凡象皆氣也'.

177. It is not quite accurate to translate Shen as 'spirit', since, according to Zhangzai, it means a subtle movement of *ch'i*. See Zhang Dainian, 'Zhangzai the 11th Century Materialist', in *Zhang Dainian Collected Works 3* (Hebei: Hebei People's Publishing), 248–9.

178. '知虛空即氣,則有無、隱顯、神化、性命、通一無二'.

179. To equate the void with *ch'i* is also an attack against the concept of the void in Buddhism and Daoism.

180. Cheng Hao (程顥, 1032–1085) and Chen Yi (程頤, 1033–1107) developed a theory based on '*Li*' (reason) or more precisely '*tian Li*' (reason of the heaven), which was further developed by Zhu Xi (朱熹, 1130–1200).

181. Zhu Xi (朱熹), *Collected Works* 58. 'A Reply to Huang Dao Fu' (文集卷58·答黃道夫) (Taipei: Wu Foundation, 2000), 2799, '天地之間，有理有氣。理也者,

Without entering into the various discourses on the *Dao* and *Qi* relation among the Song Confucians, it is fair to say that *Dao* and *Qi* are distinguished but assumed to be inseparable, while *ch'i* and *Qi* are simply identified with each other. But how can *ch'i* be the equivalent to *Qi*, unless *Qi* has simply been taken as a 'natural object'?[182]

The debate on the position of *ch'i* was not resolved even up to the time of Mou Zongsan (1909–1995). Mou proposes that, for Zhang Zai, *tai he* means two things: *Ch'i* (氣) and *tai xu* (太虛, great void), which is *shen* (神, 'spirit'). Mou insists that the *Li* (理) are not sufficient to set *ch'i* into movement, since they are only principles, and therefore require a 'primary mover'. This primary force resides in *xin, shen*, and *qing* (情, 'emotion').[183] *Ch'i* (氣), *li* (理), and *xin* (心) continued to compete to be the most fundamental metaphysical principle of Neo-Confucianism, with philosophers trying either to integrate them or to argue for one over the other. For Mou, *xin* stands out as the strongest candidate. Yet how do these subjective forces drive being into movement? Mou has no other way of explaining this apart from taking a Kantian stance, where the trinity (*ch'i, li, xin*) is the condition of possibility of the experience of phenomena, and existence and experience are correlated. Another key philosopher, Zhang Dainian (張岱年, 1909–2004) held a different view, and radically interpreted Zhang Zai as an eleventh-century materialist—not an unreasonable proposition

---

形而上之道也，生物之本也。氣也者，形而下之器也，生物之具也'.

182. Zhang Dainian (張岱年) confirms that Zhang Zai understands *ch'i hua* as *Dao*, and the Cheng brothers' *li* as *Dao*, the *Dao-Qi* question is transformed to the question of *li-ch'i*—meaning that the *Qi* question was obscured. 'Analysis of the Li-Ch'i Question in Chinese Philosophy [中國哲學中理氣事理問題辯析]', *Chinese Cultural Studies* (中國文化研究) 1 (2000), 19-22: 20

183. Mou Zongsan, *Lectures On the Philosophy of Zhou Yi* (周易哲學演讲录,) (Shanghai: East China Normal University Press, 2004), 59.

given that Zhangzai himself says that '*tai xu* is *ch'i*' (太虛即氣)—that the force is present in *ch'i*, but not external to it.[184] This debate deserves more detailed study than we can offer here. However, it is difficult to accept either Zhang Dainian's materialist argument or Mou Zongsan's 'primary mover' argument, since both of them seem inadequate to fully account for the role of *Qi*. They seek a 'prime mover' either in matter or in the spirit.[185] Even if we want to describe Song's thought as a kind of materialism, it must be said that his concept of *ch'i* is not a substantialist materialism, but rather a relational materialism. In Song, the monism of *ch'i* is developed into five elements: metal, wood, water, fire, and earth, each with a unique composition of *ch'i*. This resonates with Presocratic thinking, and yet is fundamentally different. These five elements are called *Wu Xing* (五行)—literally, 'movements'; they are not *substantial* elements, but *relational* movements. Song takes up Zhang's *ch'i*, and, in his 'On *ch'i*', proposes that 'what fills out heaven and earth is *ch'i* (盈天地皆氣也)'.[186] He continues:

---

184. Mou Zongsan argued that Zhang Zai should not be taken as a monist of *ch'i*: See Mou Zongsan, *Collected Works* 5, *Moral Creative Reality: Mind and Nature* vol. 1, 493. Mou says that this misreading of Zhangzai in the work of the Cheng brothers, and then in Zhuxi, led to the incorrect conclusion that Zhangzai had proposed a monism of *Qi*, and that this reading should be corrected. '橫渠於《太和篇》一則云：『散殊而可象為氣，清通而不可象為神。』再則雲：『太虛無形，氣之本體』。復雲：『知虛空即氣，則有無、隱顯、神化、性命通一無二。』又雲：『知太虛即氣，則無無。』凡此皆明虛不離氣、即氣見神。此本是體用不二之論,既超越亦內在之圓融之論。然圓融之極,常不能令人元滯窒之誤解,而橫渠之措辭亦常不能無令人生誤解之滯辭。當時有二程之誤解,稍後有朱子之起誤解,而近人誤解為唯氣論。然細會其意,並觀諸儒家天道性命之至論。橫渠決非唯氣論,亦非誤以形而下為形而上者。誤解自是誤解,故須善會以定之也。'

185. Zhang Dainian sometimes takes a dogmatic Marxist position, arguing that Zhangzai is not materialist enough: see Zhang Dainian, *Collected Works 3*, 251.

186. '天地間非形即氣[…]由氣而化形，形復返於氣，百姓日習而不知也初由氣化形人見之，卒由形化氣人不見者，草木與生人、禽獸、蟲魚之類是也。'. Pan Jixing (潘吉星), *Critical Biography of Sung Yingxing* (宋應星評傳) (Nanjing: Nanjing University Press, 1990), 338.

> Between heaven and earth, there is either *xing* [form] or *ch'i* [...]
> *ch'i* transforms into *xing*, *xing* returns back to *ch'i*, however we
> are not conscious of it [...] when *ch'i* first comes into *xing*, we see
> it; when it returns to *ch'i* from *xing*, we don't see it.[187]

Here the individuation of beings is a transformation of *ch'i*,
from its formlessness to a concrete form—which can also
be *Qi*. Song Yingxing reformulates *Wu Xing* into a new composition, in which only earth, metal, and wood are related to
forms. Fire and water are the two most elementary forces,
situated in-between form and *ch'i*.[188] All individuated beings
in the universe are phenomena of the transformation of *ch'i*
into the forms of *Wu Xing*. These transformations also follow
the cycle of movement: when wood is burnt, it returns to the
soil. In Song's analysis, not unlike Zhang Zai,[189] he doesn't see
Wu Xing in terms of opposing forces, as in ancient philosophy
(e.g. water is opposed to fire, metal is opposed to wood), but
considers them in terms of intensities which can be combined
to produce different compositions. One might say that there is
no opposition here, but only different proportions or relations.
But for these combinations to be possible demands human

---

187. Ibid., 339.

188. Ibid., 340, 雜於形與氣之間者水火是也。

189. Zhang Zai, *Collected Works* (張錫琛點校:《張載集》) (Beijing:
Zhonghua Books, 1978), 13, Zhang describes a new dynamic of *Wu Xin* based
on intensity:「木曰曲直」，能既曲而反申也;「金曰從革」，一從革而不
能自反也。水　　火，氣也，故炎上潤下與陰陽升降，土不得而制焉。木金
者，土之華實也，其性有水火之雜，故木之為物，水漬則生，火然而不離
也，蓋得土之浮華　於水火之交也。金之為物，得火之精於土之燥，得水之
精於土之濡，故水　火相待而不相害，鑠之反流而不耗，蓋得土之精實於水
火之際也。土者，　物之所以成始而成終也，地之質也，化之終也，水火之
所以升降，物兼體　而不遺者也。'

intervention, and this is where *Qi* (器) comes in. *Qi* or technics is what brings *ch'i* into forms which may not spontaneously occur in themselves. This is a technical dimension of *Qi* which the New Confucians and the Neo-Confucians disregard when they see the heart (*xin*) or the principle (*li*) as the sole 'primary mover' of the causality of phenomena. Song is very precise on this point in 'On *ch'i*'. His argument can be summarised in two points: firstly, *ch'i* can take on forms such as water and fire, and although these elements are opposed to each other, they actually share a common attraction to one another. He uses the metaphor that, when they don't see each other, they miss each other like wife and husband, mother and son. However, they can 'see each other' through human interventions—more precisely, technical activities. Secondly, if we consider a glass of water and a chariot made of wood, when the wood is set on fire, the glass of water cannot produce any effect and will be vaporised by the fire; however, if there is a huge container of water, then the fire will easily be extinguished. Hence it is the question of *intensity* rather than that of *substance* that is essential to technological thinking.[190] Retrospectively, these thoughts can be seen to be present in the technical descriptions of Song's encyclopaedia *Tian Gong Kai Wu*. For example, in the section on clay-making Song writes that 'when water and fire are in good proportion, earth can be firmly combined to be clay or porcelain';[191] and in the section on metallurgy, both fire and water are the necessary condition for iron: 'When iron is heated and forged, it doesn't yet acquire the quality since the correct proportion between

---

190. Pang Jixing, *Critical Biography of Sung Yingxing*, 353.

191. 天工開物·陶埏, '水火即濟而土和'.

water and fire is not yet reached: when one takes it out, and sprinkles with water, then it is hard iron.'[192]

Thus *ch'i*, according to the principles of *Dao*, is actualised in different elementary movements; and through human intervention, they are reactualised so as to yield individuated beings—for example in forging, and more generally in the production and reproduction of *Qi* (器). *Qi* thus enters into the circle and enlarges the possibility of combinations of the elementary forms. We might say that the dominant philosophy of nature guided technological thinking in such a way that the artificial had always to be subsumed not only under the principles of movement that we would call physics today, but also under an organic model of combination, a mediation of the relations between different individuated beings. One has to add here that, like Liu Zongyuan, Song was sceptical of the theory of correlation between the Heaven and the human, regarding it as superstition. In his 'On Heaven' he mocked the ancients—including the descriptions of the *Classic of Poetry* and *Zuo Zhuan* discussed above (§9) and the Neo-Confucian and commentator on *Classic of Poetry* Zhu Xi (1130–1200)—of not understanding the Heaven,[193] since if the solar eclipse is correlated with moral conduct of the emperor, then any exceptions to this correlative rule will remain inexplicable. For Song, the virtue of the emperor is indicated not by such natural phenomena, but by his ability to understand the Heaven

---

192. '凡熟鐵、鋼鐵已經爐錘，水火未濟，其質未堅，乘其出火之日，入清水淬之，名曰健鋼、健鐵'.

193. Song Yingxing, 'On Heaven', <http://ctext.org/wiki.pl?if=gb&chapter=527608>, '朱注以王者政修，月常避日，日當食而不食，其視月也太儌。《左傳》以魯君、衛卿之死應日食之交，其視日也太細。《春秋》：日有食之。太旨為明時治歷之源。《小雅》：亦孔之醜。詩人之拘泥於天官也。儒者言事應以日食為天變之大者，臣子儆君，無已之愛也.'

according to 'scientific principles' in order to act in time.[194] This is to say that, even though Song questioned the theory of resonance,[195] he nevertheless confirmed the unity between the cosmos and the moral.[196]

To sum up, in terms of what we characterised earlier as cosmotechnics, we have seen the use of *Li Qi* in Confucianism to consolidate the cosmological and moral order; in Zhuangzi's case, the 'using' or 'not using' (but not usage according to its technical and social determination) of the tool to mediate with *Dao* to acquire the art of living; whereas in Song's work we see instead its role in both creating and using, in which the *Qi-Dao* moral relation is extended into everyday production. This organic form is not what we understand today in a biological sense, but finds its highest principle in *Dao*—a cosmotechnics that binds the human to the cosmos.

§14. ZHANG XUECHENG AND THE HISTORICISATION OF DAO

During the Qing dynasty (1644–1912) the relation between *Dao* and *Qi* was reformulated in a yet another way, which

---

194. Ibid. '而大君征誅揖讓之所為，時至則行，時窮則止。與時污隆，乾坤乃理。此日月之情，天地之道也。'

195. Pang Jixing, in his *Critical Biography of Song Yingxing* (363–368), claims that Song rejected the theory of resonance; however, in his text Song does not even mention the word 'resonance', rather he was rejecting the superficial correlation between the solar eclipse and evil.

196. Y. S. Kim claims that there is a 'natural theology' in Song's work since the *tian* is considered as 'creator of things'. The argument does not hold, however, since Kim ignores Song's close relation to the Neo-confucianism. Y.S. Kim, '"Natural Theology of Industry" in Seventeenth-Century China?: Ideas about the Role of Heaven in Production Techniques in Song Yingxing's *Heaven's Work in Opening Things* (*Tiangong kaiwu*)', in J.Z. Buchwald (ed.), *A Master of Science History: Essays in Honor of Charles Coulston Gillispie* (Dordrecht: Springer, 2012), 197–214.

anticipated the rupture that followed the Opium Wars. One should not, however, have the impression that the thinkers of this era deliberately wanted to break the unity of *Qi* and *Dao*; on the contrary, they attempted to reaffirm it. However, living as they did through a critical historical period, they were forced to integrate Western thought and technology into a singular philosophical system which could not be made wholly compatible with them. In order to integrate them in a 'coherent' way, they could only twist the meanings of both so as to minimise this incompatibility.

During the mid- to late Qing dynasty, it is to be noted that the aim of studying the six classics, namely the *Classic of Poetry* (詩經), the *Book of History* (尚書), the *Book of Rites* (禮記), the *Book of Changes* 周易, *Spring and Autumn Annals* (春秋), and the lost *Book of Music* (樂經), was also challenged. If, in the past, the study of the classics had focused on the philosophical analysis, textual analysis, and philological study (訓詁學) of these ancient texts in order to understand *Dao*[197] — which was given a moral meaning, *de* (遵德性) — during the Qing dynasty we see an effort toward the *historicisation* of such an understanding of *Dao* (道問學). This is a significant change in the history of thought in China, since it challenged the conception of *Dao* as something which had been already pronounced and had since then been latent in the ancient texts, and proposed instead that what is conceived to be *Dao* is historical — that *Dao* changes over time. Zhang Xuecheng (章學誠, 1738–1801) — something like the Michel Foucault of eighteenth-century China — consistently showed that one should study *Dao* in terms of meanings contextualised in time

---

197. D.S. Nivison, *The Life and Thought of Chang Hsüeh-ch'eng* (Stanford, CA: Stanford University Press, 1966), 152.

and space. Zhang opened his magnum opus *On Literature and History* (文史通义) with the following statement:

149

ZHANG XUECHENG AND THE HISTORICISATION OF DAO

> The six classics are only histories. The ancients didn't write books without a purpose in mind; the ancients never theorized without basing their work on facts, and the six classics are the political manuals of the ancient emperors.[198]

Zhang's statement differentiates him from his contemporary, the famous Confucian Dai Zhen (戴震, 1724–1777), well known for his philological research. Dai was very critical of Neo-Confucianism, especially its interpretation of *Li* (理, 'reason'). He famously denounced Zhu Xi and other later Confucians for 'using *Li* for killing', just as cruel officials use the law for deadly ends.[199] For Zhang, Dai was still trapped in the tradition which seeks the *Dao* in the ancient texts, not realising that the six classics cannot transcend time—and that, if they could, that would mean that the *Dao* would become eternal, which would itself be a contradiction. For Zhang, the six classics tell us about nothing more than the *Dao* of their time. For the inquiry into the *Dao* of our own time, it is necessary to historicise according to the development of society and the complications this development brings with it. This historicisation is also a philosophising, which takes a leap into the philosophy of history instead of lingering over the endless decoding of 'original meanings'. Moving away from a detailed

---

198. '六經皆史也。古人不著書；古人未嘗離事而言理，六經皆先王之政典也.'. Cited by Yu Ying-shih (余英時), *On Dai Zhen and Zhang Xuecheng: Research of the History of Thought in the Mid Qing Dynasty* (論戴震與章學誠：清代中期學術思想史研究) (Beijing: SDX Joint Publishing Company, 2000), 57.

199. Chen Lai, *Sung Ming Li Xue*, 6.

analysis of etymology, Zhang therefore suggests philosophising about history in a more general way, in an approach that his biographer David Nivison sees as comparable to Hegel's analysis of history.[200] The six classics thus become the *Qi* of the *Dao* of antiquity. In the chapter '*Yuan Dao Zong*' (《原道中》, 'Original Dao'), Zhang states:

> The *I Ching* says: 'what is above the form is called *Dao*, what is below the form is called *Qi*'. *Dao* and *Qi* are not separable, just as shadow cannot be detached from shape. The later scholars believed that the teaching of Confucius comes from the six classics, and thought that the six classics are where *Dao* dwells, without knowing that the six classics are actually only *Qi* [...] Confucius passed six classics to the next generation, since the *Dao* of the ancient emperors and sages are not visible without the six classics acting as *Qi* [...] The Confucians kept on believing that the six classics contain *Dao*; but how can one talk about *Dao* without *Qi*?—how can shadow exist without shape?[201]

In a certain sense, what Zhang Xuecheng suggests here is close to what we call deconstruction today: here, the existence of *Dao* also depends on its *supplement*—its *subjectile* as Derrida might say—otherwise it would become invisible. Writing, in particular here the writing of history, is significant in that it renders visible the *Dao* that continually changes and

---

200. Nivison, *The Life and Thought of Chang Hsüeh-ch'eng*, 158.

201. '《易》曰：「形而上者謂之道，形而下者謂之器。」道不離器，猶影不離形，後世服夫子之教者自六經，以謂六經載道之書也，而不知六經皆器也。……夫子述六經以訓後世，亦謂先聖先王之道不可見，六經即其器之可見者。……而儒家者流，守其六籍，以為是特載道之書；夫下豈有離器言道，離形存影者哉！'. Cited by Yu Yingshih, *On Dai Zhen and Zhang Xue Cheng*, 53;

moves above visible forms. It is clear that Zhang confirms the unity of *Dao* and *Qi*; but paradoxically, in doing so, he also relativises the relation between *Qi* and *Dao*, making it a historical phenomenon. Zhang's conception of *Dao* and *Qi* was to have a great influence on scholars such as Gong Zizhen (龔自珍, 1792–1841) and Wei Yuen (魏源, 1795–1856), important figures in the early modernisation of China, discussed below.[202] Zhang's critique of Neo-Confucianism also shifted the focus from the correlation between knowing and morality toward knowing and objective knowledge—this point, although implicit, would be important for the programme of New Confucianism.[203]

§15. THE RUPTURE OF QI AND DAO AFTER THE OPIUM WARS

Under fierce attack for its metaphysical discourses, considered to be empty and detached from history and reality, Neo-Confucianism fell into decline from the end of the Ming dynasty and finally gave way to the new disciplines of Western science towards the end of the Qing dynasty. This is a development that is far more difficult to account for than the spread of Buddhism during the Wei Jin dynasties. Buddhism brought about a new form of thinking and new values, but the existing values and powerful material supports embedded in Western science would make it impossible for the latter to be immediately

---

202. Yu Ying-shih (余英時), *The Humanity and Rationality of China* (《人文與理性的中國》) (Taipei: Linkingbooks, 1998), 395.

203. Yu Ying-shih, *On Dai Zhen and Zhang Xue Cheng*, 89–90. Yu points out three main differences between Zhang and Wang Yangmin, whose concept of *Liangzhi* is fundamental to Mou Zongsan's philosophical programme, discussed in §18 below. The main difference could be interpreted as the shift from what Wang calls the *Liangzhi* (knowing good, or the heart) of *de* (virtue) to the *Liangzhi* of knowledge.

adopted. Instead, it enforced an *adaptation* to the technologi-cal condition. This adaptation represents one of the greatest challenges and crises that Chinese civilisation has experienced, and indeed seems to render impossible any return to a 'proper', 'authentic' origin.

Western technology produced hype in China; but more fundamentally, it produced fear. Take the example of the first railway in China, from Shanghai to Woosung, built by the British company Jardine, Matheson & Co. in 1876–1877. The railway provoked such anxiety (in terms of security and potential accidents) that the Qing dynasty paid 285,000 tales of silver to buy the railway and subsequently destroyed it.[204] The cultural transformation in question here, which some Asian scholars ambiguously tend to call 'a different modernity', is indeed very modern in the sense that it is remarkably 'Cartesian': for the attempt to impose scientific and technological development while retaining the 'fundamental principles' of Chinese thought implies that the mind (the *cogito*—or, here, philosophical thought), through the medium of technics, can contemplate and command the physical world without itself being affected and transformed.

The two Opium Wars in the mid-nineteenth century had destroyed the civilisation's self-confidence, and thrown it into a whirlpool of confusion and doubt. After the Opium Wars (1839–1842, 1856–1860), China recognised that it would be impossible to win any war without developing 'Western' technologies. The serious defeats it suffered led to the Self-Strengthening Movement (自強運動, 1861–1895), which

---

204. Sun Kuang-Teh (孫廣德), *Late Ching Tradition and Debates Around Westernisation* (晚清傳統與西化的爭論) (Taipei: Taiwan Commercial Press, 1995), 29.

extensively modernised the military, industrialised production, and reformed the education system. Two of the slogans of this movement capture the spirit of the times. The first is 'learning from the West to overcome the West [師夷長技以制夷]'; the second bespeaks a more cultural and nationalist spirit: 'Chinese learning for fundamental principles and Western learning for practical application [中學為體，西學為用]'. Li Sanhu has pointed out that, in the confrontation between Chinese and Western culture, a series of 'translations' took place in which *Dao* and *Qi* were gradually identified respectively with Western (social, political and scientific) theory and technology.[205] Li proposed that if, since the Han dynasty, *Dao* was understood to be prior to *Qi*, from the time of the later Ming and Qing dynasties, this order was reversed, with *Qi* being seen as prior to *Dao*.[206]

The first translation consists in replacing *Qi* with Western technology, and using it to realise the Chinese *Dao*. During the reform movement following the Opium Wars, Wei Yuan, the intellectual who proposed the slogan 'learning from the West to overcome the West', identified Western technology with *Qi* in the hope of integrating it into the traditional study of the classics. Wei had heavily criticised the Neo-Confucians for speculating on metaphysics, and not *using Dao* properly to solve social and political problems. He sought to retrieve some principles from Chinese philosophy that he thought might help reform Chinese culture from within, and accordingly read the six classics as books on governance.[207] He thus unconsciously reversed the

---

205. Li Sanhu, *Reiterating Tradition*, 111.

206. Ibid., 67. This is particularly clear in the work of Wang Fuzhi (王夫之, 1619–1692).

207. Chen Qitai (陳其泰) and Liu Lanxiao (劉蘭肖), *A Critical Biography of Wei Yuan* (魏源評傳) (Nanjing: Nanjing University Press, 2005), 159.

holistic view of *Dao* and *Qi* into a kind of Cartesian dualism. In comparison with Zhang Xuecheng, who influenced him, Wei Yuan extends the concept of *Qi* from historical writings to artefacts, and takes a much more radical materialist stance. If the Gu Wen movement was an attempt to reassert *Dao* through writing, nonetheless it still held that one could find the unity of *Dao* and *Qi*. In Wei Yuan's extension of the concept of *Qi* to Western technologies, he breaks definitively with the moral cosmology: *Qi* becomes a mere thing controlled and mastered by *Dao*. *Dao* is mind, and *Qi* is its instrument. In this conception, *Qi* becomes a pure tool. Yan Fu (嚴復, 1894–1921), translator of Thomas Huxley and Charles Darwin, mocked this 'matching' between the Chinese *Dao* and the Western *Qi*:

> The body and its use reside within a unity. The body of a cow is used to carry loads; the body of a horse to voyage. I have never heard that the body of a cow can be used like a horse. The difference between the East and the West is like that between two different faces; we cannot ignore this and claim that they look similar. Therefore, Chinese thought has its own use, and so does Western thought; they should be juxtaposed, and when they are unified, they will both perish. Those who want to combine them as one thing, while separating them one part as body and one part as instrument, already commit a logical error; how can we expect this to work?[208]

---

208. '體用者，即一物而言之也。有牛之體，則有負重之用；有馬之體，則有致遠之用。未聞以牛為體，以馬為用者也。中西學之為異也，如其種人之面目然，不可強為似也。故中學有中學之體用，西學有西學之體用，分之則並列，合之則兩亡。議者必欲合之而以為一物，且一體而一用之，斯其文義違舛，固已名之不可言矣，烏望言之而可行乎？（《嚴復集》第三冊，1986年：558–9）', Li Sanhu, *Reiterating Tradition*, 109,

The second translation, according to Li Sanhu, consists in replacing both *Dao* and *Qi* with Western theory and Western technology. What followed the Self-Strengthening Movement (洋務運動, 1861–1895) was the Hundred Days' Reform (戊戌維新, 11 June–21 September 1898), a reaction from intellectuals to the shock of China having been defeated by Japan in the First Sino-Japanese War (1894–1895). Retrospectively, we can well imagine why this event was registered as a trauma: defeat by the Western countries could be explained by the relative advancement of their civilisation, whereas defeat by the Japanese must have seemed inexplicable, considering that Japan was a small 'subordinated state' of China. After the Opium Wars, the Self-Strengthening Movement aimed firstly to strengthen the military in China, developing better quality warships and weapons; and secondly to integrate Western sciences and technology into China through industrialisation, education, and translation. However, all of these plans were suspended due to the defeat in the Sino-Japanese War.

We must note that, at this moment, materialist thought was rather popular in Europe, and Chinese intellectuals who had become more familiar with European thought started to appropriate it. Let us take one of the most famous reformist intellectuals, Tan Sitong (譚嗣同, 1865–1898), as an example. Like almost all Confucians, Tan also emphasised the unity of *Dao* and *Qi*. However, as Li Sanhu pointed out, he equated *Qi* with science and technology, and identified *Dao* with Western scientific knowledge, albeit formulated in terms of Chinese philosophical categories. His materialist thinking holds that *Qi* is the support of *Dao*; without *Qi*, *Dao* would no longer exist. Therefore *Dao* should be changed to be compatible with the Western '*Qi*'. Consequently, Tan effectively reversed Wei Yuan's

'*Qi* in the service of *Dao*' (器為道用) into '*Dao* in the service of *Qi*' (道為器用).

The 'materialists' of this period combine Chinese philosophy with Western science in very creative, sometimes seemingly absurd ways. In Shanghai in 1896, Tan met the English Jesuit John Freyer (傅蘭雅, 1839–1928), who introduced the concept of the ether into China.[209] Tan performed a materialist reading of the ether, and translated it into his earlier reading of the Chinese classics, including the *I Ching* and Neo-Confucianist texts. He proposed to understand Confucius's *ren* (仁, 'benevolence') as the 'use' or 'expression' of the ether:

> Of the *Dharma* world, the spiritual world, the world of sentient beings, there is a sublime being that sticks to all, unites all, and channels all, fills all. It can't be seen, can't be heard, can't be smelled, can't be named, and we call it the ether. It gives birth to the material world, the spirit, and the sentient beings.
>
> *Ren* (benevolence) is the use of the ether, all beings in the universe come from there, and communicate through it.[210]

Observe that, since *ren* is the spiritual part of the ether, the ether is both *ch'i* and *Qi*, and *ren* is its *Dao*. Retrospectively, we may suspect that Tan had indeed found a substitute for the Neo-Confucian's *ch'i* in the ether, and hence wanted to realise the *Dao* through the study of the ether. At the time Tan was also reading John Freyer's translation of Henry Wood's *Ideal Suggestion Through Mental Photography* (translated

209. Ibid., 113.

210. '徧法界、虛空界、眾生界,有至大之精微,無所不膠粘、不貫洽、不筦絡、　而充滿之一物焉,目不得而色,耳不得而聲,口鼻不得而臭味,無以名之,名之曰「以太」。……法界由是生,虛空由是立,眾生由是出。　夫仁,以太之用,而天地萬物由之以生,由之以通'.

into Chinese by Freyer as 治心免病法, which can be trans-lated back as: *Ways to Get Rid of Psychological Illness*), in which the author suggests an analogy between the move-ment of the ripple and the psychological force.[211] This image matches perfectly Tan's speculation on the relation between the ether and *ren*, which he developed into what he called a 'theory of the force of the heart', or 'psychological power (心力說)'.

Tan's comrade, another famous reformist intellectual, Kang YouWei (康有為, 1858–1927), put forward a similar interpreta-tion, stating that '*ren* is thermal force; *yi* (justice, rightfulness) is gravitational force; there is no third that constitutes the universe'.[212] From the outset, we can understand them (among with other similar theories came out of that period) as attempts to re-unify *Qi* and *Dao*; however, the mismatch of categories and their meanings produced incompatible mixtures which couldn't help but end in failure.

This absorption by Chinese intellectuals of nineteenth-cen-tury physics as a new foundation for Chinese moral philosophy so as to boost popular hopes of realising political and social equality is but a particularly poignant example of the type of appropriation with which intellectuals sought to reinvigorate Chinese thought with Western science and technology. In 1905, after eight years of exile in the United States and Europe, Kang wrote a book entitled *Saving the Country through Material* (物質救國論), in which he states that the weakness of China is

---

211.   Bai Zhengyong (白崢勇). On the main concerns in Tan Sitong's thought, see 'Exploring Tan's Thought on Ether, Benevolence and Psychic Power' (從「以太」、「仁」與「心力」論譚嗣同思想之旨趣), 文與哲 12 (2008), 631–2.

212.   Li Sanhu, *Reiterating Tradition*, 112, 「仁者，熱力也；義者，重力也；天下不能出此兩者」(《康子內外篇　人我篇》).

not a matter of morality and philosophy, but rather of material; the only way to save China, then, is to develop a 'science of material' (物質學).[213] What Kang means by 'material' is actually technology.[214] This reading is perfectly compatible with a movement of modernisation, understood as the making use of '*Qi*' in order to realise '*Dao*'. This instrumentalist emphasis on 'usage' or 'use' reverses the *Dao* and *Qi* of cosmotechnics—and, according to Li Sanhu, replaces Chinese holism with Western mechanicism.[215]

§16. THE COLLAPSE OF QI-DAO

A second major period of reflection on science and technology, as well as democracy, came after the 1911 revolution in China, when some of those who had been sent abroad as children later returned as public intellectuals. One of the most important intellectual movements, now known as the May Fourth Movement, erupted in 1919, initiated by the protest against the Treaty of Versailles which allowed Japan to take over some territories in Shang Dong province, previously occupied by the Germans. More significantly, it also led to a movement among a young generation who concerned themselves not only with science and technology, but also with culture and values. On the one hand, this cultural movement resisted traditional authorities; on the other hand it placed a high value

---

213. Lou Zhitian (羅志田), *Tradition in Disintegration. Chinese Culture and Scholarship in Early 20th Century* (裂變中的傳承 - 20世紀前期的中國文化與學術) (Beijing: Zhonghua Book Company, 2009), 328.

214. Ibid., 331. However on page 219 of the same book, Lou says that it refers to science; we can see that science and technology are two concepts that were not, and still are not, well distinguished among scholars in China.

215. Ibid. The notable example that Li gave is the anarchist Wu Zhihui (吳稚暉, 1865–1953), founder of the Institut Franco-Chinois de Lyon, who promoted mechanism as utopia.

on democracy and science (which were popularly known as 'Mr De and Mr Sai'). During the 1920s and '30s, Western philosophy started to flourish in China.

Three names are closely related to the contemporary intellectual history of China: William James, Henri Bergson, and Bertrand Russell.[216] The intellectual debates of the period concerned whether or not China should be fully Westernised and fully adopt Western science, technologies, and democracy—as advocated by intellectuals such as Hu Shi (a student of John Dewey), and, on the opposite side, criticised by Carsun Chang Junmai (a student of Rudolf Eucken), Chang Tungsun (the Chinese translator of Bergson in the 1920s), and others. These debates, however, led to unresolved questions and uncompromising propositions. The question raised at this time, one that anticipated the advent of the New Confucianism, was how it might be possible to develop a modernisation that would be authentically Chinese. In the following I recount some historical episodes indicative of how intellectuals of the time understood this question, and how they thought about the development of China in relation to science and technology.

§16.1. CARSUN CHANG: SCIENCE AND THE PROBLEM OF LIFE

The first episode takes place in 1923, when the philosopher Carsun Chang (張君勱, 1887–1968), an expert in Neo-Confucianism and a student and collaborator of Rudolf Eucken, delivered a talk at Chinghua University in Beijing, and later published it as an article entitled 'Rensheng Guan (人生觀)'. The title is difficult to translate: literally it means the intuition of

---

216. It is worth noting that none of these philosophers were specialists in technics, with the arguable exception of Bergson in his *Creative Evolution*.

life, or of living, and we may suppose that it was intended to evoke the German word *Lebensanschauung* used by Eucken. Chang met the latter in 1921 in Jena, and decided to study under him, later collaborating with him on a book titled *Das Lebensproblem in China und in Europa* (*The Problem of Life in China and in Europe*, 1922), which was never translated into Chinese (nor into English).[217] The book is divided into two parts: the first on Europe, by Eucken, and the second on China, by Chang, with a closing epilogue by Eucken. It is not in itself a particularly profound investigation into the subject matter, and consists of brief sketches of the different *Lebensanschauungen* ('life-views' or 'views of life') from ancient times up to the present day. In the epilogue Eucken made the following remarks on the Chinese way of life and its relation to Confucian moral philosophy:

> What we specifically found there is a strong concern with the human and his self-awareness; the greatness of this way of life lies in its simplicity and its truthfulness; in a strange way, the high esteem of such social and historical being-together was combined with rational enlightenment.[218]

Evidently, the collaboration with Eucken allowed Chang to combine his moral philosophy with the question of life. In this regard Chang calls himself a 'realist idealist', meaning that he

---

217. R. Eucken and C. Chang, *Das Lebensproblem in China und in Europa* (Leipzig: Quelle und Meyer, 1922).

218. Ibid., 199. 'Als eigentümlich fanden wir dabei namentlich die Konzentration des Strebens auf den Menschen und auf seine Selbsterkenntnis; die Größe dieser Lebensgestaltung liegt in ihrer Schlichtheit und ihrer Wahrhaftigkeit; In merkwürdiger Weise verband sich hier mit vernünftiger Aufklärung eine große Hochschätzung des gesellschaftlichen und geschichtlichen Zusammenseins.'

starts with the 'I', but that the 'I' is not posited as absolute, since it is exposed to the experience of the real world. The idealist starting-point characterises his *Lebensanschauung*, which differentiates objective science from philosophy. In *'Rensheng Guan'* Chang suggests that the 'I' should provide the perspective from which to understand that which falls outside of it, including the individual, the social, property— from the inner spiritual self to the outer material world, the hopes of the world, and even the creator. For Chang, science, a discipline that begins and ends with objectivity, should have its basis in the intuitive and subjective 'I'. Chang characterised the differences between science and the 'vision of life' in five points:

Science (based on)	Vision of life (based on)
Objective	Subjective
Reason	Intuition
Analytic method	Synthetic method
Causality	Free will
Commonality	Singularity

This schematic distinction was immediately attacked by geologist Ding Wenjiang (丁文江, 1887–1936), who criticised Chang for regressing from science to metaphysics, and dubbed his philosophy *Xuan Xue* (玄學), a term used to describe the philosophy that emerged during the Wei Jin dynasties, greatly influenced by Daoism and Buddhism (see §9 above), and was generally seen as a hybrid of a scholarly discipline and a superstition.

What is most significant in this episode is Chang's concern that science was being valued over the traditional theory of knowledge in Chinese society, implying a reconfiguration of all forms of values and beliefs, including the *Lebensanschauung*. As he warns, and as Ding's critique seems to confirm, at this time in China science was in danger of becoming the ultimate measure of all forms of knowledge and, in doing so, of filtering out anything that was considered insufficiently scientific, except those elements that it considered to be harmless and merely decorative.

§16.2. THE MANIFESTO FOR A CHINA-ORIENTED CULTURAL DEVELOPMENT, AND ITS CRITICS

Another episode from 1935 characterises the second moment and the debates around it, and allows us to understand the main ideas at stake. On 10 January 1935, ten well-known professors in China published an article entitled 'A Manifesto for a China-Oriented Cultural Development (中國本位的文化建設宣言)',[219] in which they criticised the proposal for 'Chinese thought as body and Western thought as instrument' as superficial, and demanded deeper reform. They also criticised the proposal for full westernisation, whether it be in imitation of Britain and the USA, the Soviet Union, or Italy and Germany. This manifesto expressed a fear of a chaotic internecine intellectual war that would lead to a forgetting of Chinese origins and contemporaneity alike, and envisioned a new China capable of effectively integrating technology and science without losing its roots. On 31 March, Hu Shi responded mockingly to the manifesto, claiming that there was no need to worry about a 'Chinese-oriented culture', since China will always be China. According to Hu, there is

---

219. Wang Xin Ming et al, 'A Manifesto for a China-Oriented Cultural Development', in *Westernisation/Modernisation*, vol. 2 (從西化到現代化·中冊) (Hefei: Huangshan Books, 2008).

a kind of inertia in culture in general, so that when Chinese culture attempts to fully westernise itself, it will always create something else owing to this inertia: 'Even if the Chinese accepts Christianity, as time goes on, he will be different from a European Christian, he will be a Chinese Christian'. He goes on to mock the then leader of the Chinese Communist Party Chen Duxiu (陳獨秀, 1879–1942), later a Trotskyist after he was expelled from the party: 'Chen Duxiu has accepted communism, but I believe that he is a Chinese communist, different from the communists from Moscow.'[220]

This pragmatist attitude was to become the dominant view in China, probably because it was the thinking that best befitted such a period of experimentation and questioning. Yet it is also a peculiar kind of pragmatism, since it affirms westernisation while anticipating a differentiation originating from the obstructive forces of its own culture and tradition. On this view, the Chinese culture becomes purely 'functional aesthetics' in Leroi-Gourhan's sense, meaning that it serves only to add an aesthetic dimension to the main driving forces of development, which will henceforth be Occidental—science and technology, democracy and constitutionalism. During this 1935 debate, Chang Tungsun (張東蓀, 1886–1973, translator of Henri Bergson) posed a question which was not picked up on by other intellectuals, but which remains a valid and critical one: he insists that the question *is not whether westernisation is good or bad, but rather whether China has the capacity to absorb Western civilisation at all*—a question that still resounds today amidst the social, economic, and technological catastrophes befalling the country. The kind of pragmatism exemplified by Hu has the naivety to believe that differentiation is a natural

---

220. Hu Shi, editorial of *Independent Critique* (獨立評論) 142 (March 1935).

product, and is devoid of political struggle. The pragmatic view was replaced by the Marxist doctrine at the beginning of the Communist regime, but saw a revival towards the end of the twentieth century, following the economic reform in China led by Deng Xiaoping.[221] However, what is common to all phases of this process is that the spirit of the ancient cosmotechnics is fading away, and that what proves incompatible with the modern is consigned to the harmless category of 'tradition', set apart from the forces of development.

As we can see from the above two scenes from 1921 and 1935, the question of technology was rarely mentioned as such. It was rather science and democracy (or more precisely, ideology) that were central to both debates. It seemed intuitive to include technology under science, or at least to consider it as applied science. This disregard for the question of technology meant that the intellectual debates tended to remain at the level of ideology. It is not surprising to find that the scholar Wang Hui's 2008 *The Rise of Modern Chinese Thought* pays almost no attention, within its thousands of pages of well-documented materials, to the question of technology.[222] Technology merges with the question of science, and becomes invisible. Scholars of Wang's generation still confine themselves to the discourse on science and democracy; they are incapable of a more profound philosophical analysis that

221. Li Zehou (李澤厚) and Liu Zaifu (劉再復), *Farewell to Revolution* (告別革命：回望二十世紀中國) (Hong Kong: Cosmos Books, 2000). The philosopher Li Zehou proposed a 'farewell to revolution' and called for a move away from ideological debates. China, he insists, needs a new theoretical tool to manage its internal dynamics and its international relations, namely *pragmatic reason*.

222. Wang Hui, *The Rise of Modern Chinese Thought* (現代中國思想的興起) (Beijing: SDX Joint Publishing Company, 4 vols, 2008).

would take technology into account; instead, they linger over questions of 'thought', whether idealist or materialist.

§17. NEEDHAM'S QUESTION

Throughout the twentieth century, the question of why modern science was not developed in China was of continual interest to historians and philosophers. Bearing in mind once more that science has to be fundamentally distinguished from technics, this question is still germane in order for us to take the question of technics further, since the reason why modern science wasn't developed in China also explains the collapse of *Qi-Dao* in its confrontation with modernisation. Feng Youlan, a Chinese philosopher who completed his PhD thesis at Colombia University in 1923, published an article in the *International Journal of Ethics* entitled 'Why China has no Science—An Interpretation of the History and Consequences of Chinese Philosophy'. Feng was only twenty-seven years old when he published the article, but this young philosopher asserted confidently that the reason why China didn't have science is that it didn't *need* science. Feng understands science to be closely related to philosophy; or, more precisely, to be determined by certain philosophical modes of thinking. Hence, for Feng, the absence of science in China owes to the fact that Chinese philosophy prevents the scientific spirit from emerging. Feng's analysis is intriguing, although rather than really explaining the lack of science, it poses some significant questions as to the relation between science and technology, and the role of technology in China.

I will summarise Feng's argument here in a rather simplified form. Feng showed that in ancient China (during the period of Ionian and Athenian philosophy in Greece) there were nine schools, namely Confucianism (儒家), Daoism (道家), Moism

(墨家), the School of Yin-Yang (陰陽家), the School of Law (法家), the School of Logic (名家), the School of Diplomacy (縱橫家), the School of Agriculture (農家), and the Miscellaneous School (雜家). However, only the first three schools—namely Confucianism, Daoism, and Moism—were influential, and competed to become the dominant school of thought. Feng believes that Moism was the school that was closest to science, since it promoted the arts (the art of building and the art of war) and utilitarianism. Confucianism, especially through the writing of Mencius (372–289 BC), harshly disparaged Moism and Daoism; it was against Moism because of its promotion of universal love and its consequent disregard for family hierarchy, which Confucianism takes as a central value; and against Daoism because of its promotion of an order of nature, which Daoism holds to be fundamentally unintelligible.

Feng also argues that there is a certain affinity between Confucianism and Daoism in terms of their call to return to the self in order to seek moral principles. However, the nature proposed by Daoism is not a scientific and moral principle, but rather a *Dao* that cannot be named and explained, as is already announced in the first sentence of *Dao De Ching*. For Feng, the dominance of Confucianism marked the annihilation of Daoism and Moism, and hence also the annihilation of any scientific spirit in China. Even though '*ge wu* (格物)' (the study of natural phenomena so as to acquire knowledge) is fundamental to the Confucian doctrine, the 'knowledge' that it seeks is not the knowledge of the thing in question, but the 'heavenly principle (天理)' beyond the phenomenon.

Feng's analysis is very much a reductionist approach in the sense that he reduces culture to the manifestation of certain doctrines; however, it also confirms that Chinese philosophy tended to seek higher principles whose incarnation in the secular world would determine moral and political value. Furthermore, Feng fundamentally confuses science and technology, since what Moism proposed was not a scientific spirit, but rather a craftsman's spirit, exemplified in the building of houses and the invention of war machines. Thus, although Feng's account may possibly explain why technics was not a theoretical theme in ancient China, and didn't evolve into modern technology, it does not necessarily prove that there had been a *scientific* spirit before the domination of Confucianism, unless one considers that science necessarily emerges from technology. We all know today that technics continued to advance in China until the sixteenth century, at which point it was overtaken by Europe. That is to say that, even though Moism never became the dominant doctrine, technics was not annihilated; on the contrary, it flourished until the advent of what we now call European modernity.

The question posed by Feng is also that of the great historian Joseph Needham, who dedicated his lifelong project to an analysis of why modern science and technology did not emerge in China. The multiple volumes of his *Science and Civilisation in China* remain invaluable for any future development of the philosophy of technology in China. Reproaching Feng, Needham writes that the great philosopher's 'youthful pessimism' is 'unjustified'.[223] Needham showed very well that an artisanal technical culture existed in China, and that it was in many respects advanced in comparison to the same period in Europe. Given the rich materials provided by Needham and the detailed comparisons that he made, we can feel justified in setting aside Feng's conclusions, and instead appreciate

---

223. J. Needham, 'Science and China's Influence on the World', in *The Grand Titration: Science and Society in East and West* (London: Routledge, 2013), 116.

that there was indeed a technical spirit in ancient China.[224] For Needham, though, this is a rather complicated question, which he attempted to approach through detailed analyses of the role of technicians, the feudal-bureaucratic system, and philosophical, theological, and linguistic factors. Needham defended his argument against the thesis that Chinese culture emphasised practice and hence ignored theory, which is evidently incorrect when we consider that Neo-Confucianism in China achieved speculative metaphysical heights at least as great as its mediaeval European counterparts.[225] He also defended it against the thesis that pictographic writing hindered the advancement of science in China; on the contrary, he showed that Chinese writing is even more effective and expressive than alphabetical writing, i.e. that it enables the same expression with much greater brevity.[226]

§17.1. THE ORGANIC MODE OF THOUGHT AND THE LAWS OF NATURE

Needham's argument centres on both social and philosophical factors. The main social factor is that the socio-economic system in China discouraged a technical culture from developing into its modern form, since the mark of success for individuals was to enter the bureaucratic system and to become a state 'official'. The system of selection for such appointments, based on an examination of memorisation and essay writing (starting in 605 and abolished in 1905) had a tremendous influence on China, through the effects of the material of study (mainly

---

224. Ibid., 55–122.

225. Needham, 'Poverties and Triumphs of the Chinese Scientific Tradition', in *The Grand Titration*, 23.

226. Ibid., 38.

classical texts), the way of studying, the expectations of the family, and social mobility. Needham's analysis is exemplary and I will not repeat it here. My concern is more with the philosophical explanations, upon which I find myself in agreement with Needham. He claimed that the mechanical view of the world was lacking in ancient China, and that instead what dominated Chinese thought—as we have already discussed above—was an organic and holistic view:

> [T]he *philosophia perennis* of China was an organic materialism. This can be illustrated from the pronouncements of philosophers and scientific thinkers of every epoch. The mechanical view of the world simply didn't develop in Chinese thought, and the organicist view in which every phenomenon was connected with every other according to hierarchical order was universal among Chinese thinkers.[227]

This is a significant difference (if we accept Needham's distinction between mechanical and organicist), which I believe was *cosmotechnically* determinative for the different rhythms of technological development in China and in Europe: a mechanical programme capable of effectively assimilating nature and the organic form didn't exist in China, where the organic remained always the credo of thought and the principle of living and being. This organic form of nature in China, insists Needham, must be strictly distinguished from the question of nature as it was posed in the West, from the Presocratics up to the European Renaissance. In Europe, laws—both natural laws in the juridical sense and the laws of nature—come from the same root, namely the model of

---

227. Ibid., 21.

'law-giving': in the first case, 'earthly imperial law-givers', in the second a 'celestial and supreme Creator Deity', whether the Babylonian sun-god Marduk, the Christian god, or Plato's demiurge. The Romans recognised both positive laws—civil coded laws of a specific people or State, *lex legale*, and the Law of Nations (*ius gentium*) which is equivalent to natural law (*ius naturale*).[228] The Law of nations is developed to deal with non-citizens (*peregrini*), to whom citizen laws (*ius civile*) cannot be directly applied. Although Needham did not explain the connection between the Law of Nations and the Law of Nature, we can acquire an understanding of this connection from other sources: for example, Cicero extended the Stoic law of nature to social conduct: 'The universe obeys God, seas and land obey the universe, and human life is subject to the decrees of the Supreme Law';[229] they have different connotations but the same denotation.[230] Needham believes that although *ius gentium* was hardly to be found in China, there was a sort of 'law of nature' which, as we have seen already, was the moral principle of the Heaven, reigning over both human and non-human. The natural laws of early Christianity also governed both the human and the non-human, as we can see from the definition of Natural Law by the jurist Ulpian (170–223):

> Natural Law is that which Nature has taught all animals; for that kind of law is not peculiar to mankind, but is common to all animals. [...] Hence comes that union of the male and female

228. Ibid., 300.

229. Cicero, *On the Republic. On the Laws*, tr. C. W. Keyes (Cambridge, MA: Harvard University Press, 1928), 461.

230. See J. Bryce, *Studies in History and Jurisprudence* (New York: Oxford University Press, 2 vols., 1901), vol. 2, 583–6

which we call marriage; hence the procreation and bringing up
of children.[231]

A radical separation was made, as Needham suggested, by the
theologian Francisco Suárez (1548–1617).[232] Suárez proposed
a separation between the world of morality and the world
of the non-human: law can only be applied to the former,
since things lacking reason are capable neither of law nor
obedience.[233] This concept of the law of nature with a direct
relation to the law-giver is present not only in the juridical
domain, but also in natural science, for example in Roger
Bacon and Isaac Newton. Needham proceeds to the claim
that the law of nature in the sense of *ius gentium* or natural
science in Europe is not present in China, precisely because
(1) there was a distaste for abstract codified laws owing to
historical experience, (2) *li* proved to be more suitable than any
other forms of bureaucratism, and (3) more importantly, the
Supreme Being, although it existed for a short period in China,
was depersonalised, and hence a celestial supreme creator
who gives laws to both human and non-human nature never
really existed. Therefore,

> [t]he harmonious co-operation of all beings arose not from the
> orders of a superior authority external to themselves, but from

---

231. Ibid., 588n1.

232. It is perhaps no coincidence that both Heidegger and Étienne Gilson
pointed out that Suaréz played an important role in the redefinition of the
relation between existence and essence in the history of ontology; see
M. Heidegger, *The Basic Problems of Phenomenology*, tr. A. Hofstadter
(Indianapolis: Indiana University Press, 1983), 80–83, and E. Gilson, *L'être et
l'essence* (Paris: Vrin, 1972), chapter 5, 'Aux origines de l'ontologie'.

233. Needham, *The Grant Titration*, 308.

the fact that they were all parts in a hierarchy of wholes forming a cosmic and organic pattern and what they obeyed were the internal dictates of their own natures.[234]

This lack of a mechanical causal view meant that the notion of a system well-ordered according to laws did not arise; and hence China lacked any programme that sought effectively to understand beings and to manipulate them according to mechanical causalities. This mechanical paradigm could be said to be a necessary preliminary stage for the assimilation of the organic—that is, the imitation or simulation of organic operations, as for example in the technological lineage from simple automata to synthetic biology or complex systems. Needham thus poses the following analogy:

[W]ith their appreciation of relativism and the subtlety and immensity of the universe, they were groping after an Einsteinian world-picture without having laid the foundations for a Newtonian one. By that path science could not develop.[235]

There is room for doubt as to Needham's term 'organic materialism', since it is debatable whether what he is addressing here is a materialism and an organicism at all. It is perhaps more correct to say that China was governed by moral laws which were also heavenly principles; and that law, following Needham, was understood in an 'Whiteheadian organismic sense by the Neo-Confucian school'[236]—precisely what we describe here as a Chinese cosmotechnics.

---

234. Ibid., 36.
235. Ibid., 311.
236. Ibid., 325.

§18. MOU ZONGSAN'S RESPONSE

For New Confucianism, a school that emerged in the early twentieth century,[237] the question of science and technology, along with that of democracy, was unavoidable. Having recognised that the 'Cartesian' paradigm, which would seek to absorb Western development while retaining the Chinese 'mind' intact, was no more than an illusion, New Confucianism set itself the task of integrating Western culture into that of China and making it compatible with its traditional philosophical system. To put it more bluntly, the philosophers of New Confucianism sought to show, from a cultural and especially a philosophical point of view, that it is possible for Chinese thought to produce science and technology. This attempt culminated in the work of the great philosopher Mou Zongsan (1909–1995), in particular in the guise of his reading of Immanuel Kant.

§18.1. MOU ZONGSAN'S APPROPRIATION OF KANT'S INTELLECTUAL INTUITION

Mou was trained in Chinese philosophy, from the *I Ching* to Neo-Confucianism and Buddhism, as well as Western philosophy, with a certain specialisation in Kant, Whitehead, and Russell, among others. He also translated Kant's three critiques (from their existing English translations) into Chinese. Kant's philosophy plays a decisive role in bridging Western

---

237. According to Liu Shu-hsien (劉述先, 1934–)'s classification, Xiong Shili (熊十力, 1885–1968) belongs to the first group of the first generation, Feng Youlan (1895–1990) belongs to the second group of the first generation; Mou Zongsan (1909–1995) belongs to the second generation; Liu himself, Yu Yingshih (余英時, 1930–), and Tu Weiming (杜維明, 1940–) belong to the third generation. See Liu Shu-hsieng, *One Principle Many Manifestations and the Global Territorialization* (理一分殊與全球地域化) (Beijing: Peking University Press, 2015), 2.

and Chinese thought in Mou's system. Indeed, one of Mou's most striking philosophical manoeuvres is to think the division between Western and Chinese philosophy in terms of what Kant calls phenomenon and noumenon. In one of his most important books, *Phenomenon and Thing-in-Itself* (現象與物自身), Mou writes:

> According to Kant, intellectual intuition belongs only to God, but not to humans. I think this is really astonishing. I reflect on Chinese philosophy, and if one follows the thought of Kant, I think that Confucianism, Buddhism and Daoism all confirm that humans have intellectual intuition; otherwise it wouldn't be possible to become a saint, Buddha, or *Zhenren*.[238]

What exactly is this mysterious intellectual intuition that is fundamental to Mou's analysis? In the *Critique of Pure Reason*, Kant sets up a division between phenomena and noumena. Phenomena appear when the sensible data delivered through the pure intuitions of time and space are subsumed under the concepts of the understanding. But there are cases when objects that are not perceived through sensible intuition can still become objects of the understanding. In Edition A of the *Critique* we find the following clear definition:

> Appearances, in so far as they are thought as objects according to the unity of the categories, are called phenomenon. But if I assumed things that are objects merely of the understanding and that, as such, can nonetheless be given to an intuition—even if not to sensible intuition (but hence *coram*

---

238. Mou Zongsan, *Collected Works 21, Phenomenon and the Thing-in-Itself*, 5.

intuitu intellectuali)—then such things would be called noumena (*intelligibilia*).[239]

This noumenon, which sometimes Kant calls the *thing-in-itself* in Edition A, demands another, non-sensible type of intuition. The noumena as a concept is therefore negative, in so far as it poses limits to the sensible. Yet it could potentially have a positive signification if we could 'lay at [its] basis an intuition'— that is, if we could find a form of intuition for the noumenon.[240] Since such an intuition could not be a sensible one, however, it is something that human beings do not possess:

> [S]uch an intuition—viz., intellectual intuition—lies absolutely outside our cognitive power, and hence the use of the categories can likewise in no way extend beyond the boundary containing the objects of experience.[241]

Kant's refusal of intellectual intuition as something accessible to human beings is decisive for Mou's interpretation of the difference between Western and Chinese philosophy. In *Intellectual Intuition and Chinese Philosophy*, a precursor to the later and more mature *Phenomenon and Thing-in-Itself*, Mou attempted to show that intellectual intuition is fundamental to Confucianism, Daoism, and Buddhism alike. For Mou, intellectual intuition is associated with the creation (e.g. cosmogony) and with moral metaphysics (as opposed to Kant's metaphysics of morals, which is based on the subject's

---

239. I. Kant, *Critique of Pure Reason*, tr. W.S. Pluhar (Indianapolis: Hackett, 1996), A249, 312.

240. Ibid., B308, 318.

241. Ibid.

capacity for knowing). Mou finds theoretical support for this view in Zhang Zai's work, particularly in the following passage:

> The brightness of the heaven is no brighter than the sun, when one looks at it, one doesn't know how far it is from us. The sound of the heaven is no louder than the thunder, when one listens to it, one doesn't know how far it is from us. The infinity of heaven is no greater than the great void (*tai xu*), therefore the heart (*xin*) knows the heaven's boundary without exploring its limits.[242]

Mou notes that the first two sentences refer to the possibility of knowing through sensible intuitions and understanding; the last sentence, however, hints that the heart is able to know things that are not bounded by phenomena. He remarks on the strangeness of the last sentence, which is not, strictly speaking, logically meaningful, for there can be no meaningful comparison of infinities. For Mou, the capacity of the 'heart (*xin*)' to 'know the heaven's boundary' is precisely intellectual intuition: it doesn't refer to the kind of knowing determined by sensible intuitions and the understanding, but rather to a full illumination emerging from the *cheng ming* of the universal, omnipresent, and infinite moral *xin* (遍、常、一而無限的道德本心之誠明所發的圓照之知).[243] In this full illumi-

---

242. Mou Zongsan, *Intellectual Intuition and Chinese Philosophy* (智的直覺與中國哲學), 184. I adopt the translation of *tai xu* as 'great void' from Sebastian Billioud. See S. Billioud, *Thinking through Confucian Modernity: A Study of Mou Zongsan's Moral Metaphysics* (Leiden: Brill, 2011), 78, '天之明莫大於日，故有目接之，不知其幾萬里之高也。天之聲莫大於雷霆，故有耳屬之，莫知其幾萬里之遠也，天之不禦莫大於太虛，故心知廓，莫究其極也。'.

243. Mou Zongsan, *Intellectual Intuition and Chinese Philosophy*, 186.

nation, beings appear as things-in-themselves rather than as objects.[244]

Cheng ming, literally 'sincerity and intelligence', comes from the Confucian classic *Zhong Yong* ('Doctrine of the Mean').[245] According to Zhang Zai, 'the knowing of *Cheng ming* reaches the *liangzhi* of the moral of heaven, and is totally different from knowing through hearing and seeing (誠明所知乃天德良知；非聞見小知而已)'.[246] Thus knowing based on intellectual intuition characterises Chinese philosophy and its moral metaphysics. Mou often repeated that his is a moral metaphysics, but not a metaphysics of morals, since the latter is only a metaphysical exposition of the moral, while for the former, metaphysics is only possible starting with the moral. He therefore demonstrates how the unification of *Qi* and *Dao* depends upon this capacity of the mind to go beyond formality and instrumentality. Mou also demonstrates that intellectual intuition exists in both Daoism and Buddhism. It is not our purpose here to repeat his lengthy and detailed proof but, in short, intellectual intuition in Daoism is related to the fact that knowledge is infinite, while human life is finite—therefore it is futile to chase after infinity with one's limited life. We can understand this from the first two sentences of the story of Pao Ding cited above:

244. Ibid., 187.

245. In Zhong Yong, one reads 「誠者天之道也，誠之者，人之道也；自誠明，謂之性。自明誠，謂之教。誠則明矣，明則誠矣。」: 'Sincerity is the way of heaven. The attainment of sincerity is the way of men [...] When we have intelligence resulting from sincerity, this condition is to be ascribed to nature; when we have sincerity resulting from intelligence, this condition is to be ascribed to instruction. But given sincerity, we shall have intelligence; there shall be the sincerity.' tr. J. Legge, <http://www.esperer-isshoni.info/spip.php?article66>, 1893 [translation modified].

246. Mou Zongsan, *Intellectual Intuition*, 188.

> Your life has a limit, but knowledge has none. If you use what
> is limited to pursue what has no limit, you will be in danger. If
> you understand this and still strive for knowledge, you will be in
> danger for certain![247]

This would prima facie seem to confirm Kant's prohibition on intellectual intuition. But Pao Ding puts forward another way of knowing, namely that the *Dao* is that which is beyond all knowledge, and yet can be apprehended by the heart. The same is true for Buddhism, as demonstrated in the concept of the void or nothingness: the void and the phenomenon coexist, but in order to know the void, one must go beyond phenomena and physical causality.

For Anglophone readers who wish to look deeper into Mou's argument for intellectual intuition, the work of Sébastien Billioud serves as a good introduction, although Billioud also criticises Mou for remaining silent on Kant's *Critique of Judgement* and the reinterpretation of intellectual intuition in post-Kantian philosophy, especially the work of Fichte and Schelling—a reasonable enough criticism since, although he refers several times to Fichte, Mou never engages with his thought in any depth. Billioud attempts to compare Mou Zongsan and Schelling through the work of the great French connoisseur of Schelling Xavier Tilliette.[248] However, we must be careful with this comparison. The term 'intellectual intuition' is rather muddy already, and its legacy in German idealism even more so. In an influential 1981 article, Moltke Gram argued against that what he calls the 'continuity thesis' regarding intellectual intuition, as a transition from Kant to Fichte and

---

247. *Zhuangzi*, 19

248. Billioud, *Thinking through Confucian Modernity*, 81–9.

Schelling. The 'continuity thesis' comprises the following three claims summarized by Gram: (1) For Kant, intellectual intuition is a single problem; (2) the object of intellectual intuition is not given to it, but rather created by it (as for the deity); (3) Fichte and Schelling deny Kant's claim that human beings do not have intellectual intuition and affirm it as the core of their systems.[249] Gram shows that, for Kant, intellectual intuition has at least three different meanings, namely: (1) the intuition of the noumenon in the positive sense; (2) the creative intuition of an archetypal intellect; and (3) the intuition of the totality of nature. He further argues that Fichte's and Schelling's concepts of intellectual intuition basically do not correspond to any of the above three senses.[250]

In fact, if we take a close look at Fichte's and Schelling's use of the concept of intellectual intuition, we can see that it is almost opposite to Mou Zongsan's. For Fichte and Schelling, Kant's 'I think' remains a fact, a Tatsache, and so cannot furnish the ground of knowing; for the ground of knowing must be absolute, in the sense that it is not conditioned by anything else. For Fichte, beyond the 'I think', there must be an immediate consciousness of this 'I think', and it is this consciousness that has the status of intellectual intuition. In a preliminary work to the Wissenschaftslehre, 'Review of Aenisidemus', Fichte

---

249. M.S. Gram, 'Intellectual Intuition: The Continuity Thesis', Journal of the History of Ideas 42:2 (Apr–Jun 1981), 287–304.

250. Yolanda Estes responded to Gram's essay by claiming that there are actually five meanings of intellectual intuition in Kant; besides the three mentioned above, she added (4) the apperception of the I's self-activity, and (5) the conjoined intuitions of the moral law and freedom—and showed that these two meanings are affirmed by Fichte and Schelling. See Y. Estes, 'Intellectual Intuition: Reconsidering Continuity in Kant, Fichte, and Schelling', in D. Breazeale and T. Rockmore (eds), Fichte, German Idealism, and Early Romanticism (Amsterdam: Rodopi, 2010), 165–78.

claims that 'if the self of intellectual intuition *is, because* it is, and *is, what* it is; then it is insofar as it *posits itself*, absolutely self-sufficient and independent.'[251] Therefore Fichte proposes to think of intellectual intuition as *Tathandlung*, as a self-positing act. In the same way, the early Schelling understood intellectual intuition as the ground of knowing, as elaborated in his 1795 essay 'Of the I as the Principle of Philosophy'. However, there are two different developments in Fichte and Schelling, although they both face the same question of the passage from the infinite to the finite. In Fichte, the unconditional I requires a non-I as negation or as check (*Anstoß*); what is outside of the unconditional I is only the product of such a negative effect; whereas Schelling's *Naturphilosophie* moves from the I to nature, and considers that the I and nature have the same principle, as expressed in his famous claim 'Nature should be Mind made visible, Mind the invisible Nature'.[252] The Absolute, for Schelling, is no longer the subjective pole, but rather the absolute unity of subject-object, which is constantly in recursive movement. In short, it must be said that Fichte's and Schelling's concepts of intellectual intuition are based on the search for an absolute foundation of knowing, which is then turned into a recursive model, whether 'abstract materiality' in Fichte[253] or the 'productivity of nature' in Schelling.[254] This distinction

---

251.   Cited by D.E. Snow, *Schelling and the End of Idealism* (New York: SUNY Press, 1996), 45.

252.   F.W.J. Schelling, *Ideas for a Philosophy of Nature*, tr. E. E. Harris and P. Heath (Cambridge: Cambridge University Press, 1989), 43.

253.   'Abstract materiality' is a term used by Iain Hamilton Grant to describe the infinite iteration or looping of Fichte's model which explains both I and Nature; see I.H. Grant, *Philosophies of Nature after Schelling* (London: Continuum, 2008), 92.

254.   For a detailed analysis of Schelling's concept of individuation in his early Naturphilosophie, see Y. Hui, 'The Parallax of Individuation: Simondon and

between Fichte and Schelling is later described by Hegel in his *The Difference Between Fichte's and Schelling's System of Philosophy*: Fichte aims for a 'subjective subject-object', while Schelling seeks an 'objective subject-object', meaning that for Schelling nature is considered to be independent (*selbstständig*).[255] In any case, the role played by intellectual intuition in both enterprises is quite different from the use Mou intends to make of it in connecting it with the Chinese tradition.

Yet despite these differences, the inquiries of Mou certainly share something with those of the German Idealists, as far as the dynamic between the infinite and the finite is concerned. We have seen that, for the Idealists, there is a passage from the infinite to the finite, which explains being; for Mou, though, the passage leads from the finite to the infinite, since he aims not for a philosophy of nature, but a moral metaphysics. Mou Zongsan's critique of Heidegger's *Kant and the Problem of Metaphysics* rests on exactly this point: that Heidegger failed to show that Dasein is finite but can also be infinite. The ultimate difference is that Mou has no intention of finding an objective form for the inscription of the infinite in the finite, but rather seeks to found it in a formless being: *xin* (心, 'heart') as the ultimate possibility of both intellectual intuition and sensible intuition; and it is also within the infinite *xin* that the thing-in-itself can become infinite.

Mou subsequently attempts to use this division between noumenon and phenomenon to explain why there is no modern science and technology in China. In his 1962 *Philosophy of*

Schelling', *Angelaki* 21:4 (Winter 2016), 77–89.

255. B.-O. Küppers, *Natur als Organismus: Schellings frühe Naturphilosophie und ihre Bedeutung für die moderne Biologie* (Frankfurt am Main: Vittorio Klostermann, 1992), 35.

History (歷史哲學), a book that responds to Hegel's claim that China didn't have subjective freedom, Mou reiterates that Chinese philosophy has speculated about the noumenal world and paid little attention to the phenomenon and the externalisation of the spirit, which was considered to be secondary—a tendency that is expressed in various aspects of Chinese culture. Occidental culture has taken the contrary path, refraining from speculating on the noumenon and devoting itself to the phenomenon. Mou calls the former the 'synthetic spirit of comprehending reason [綜合的盡理之精神]' and the latter the 'analytic spirit of comprehending reason [分解的盡理之精神]'. In Mou's interpretation, intellectual intuition means the capacity of an intuition which is far beyond any analytic deduction or synthetic induction, and this intuition is not the sensible one which serves the understanding.[256] In other words, the intellectual intuition that Kant thought was only possible for God is also, within the framework of Daoism, Confucianism, and Buddhism, possible for human beings. The important point here, according to Mou, is that when intellectual intuition dominates thinking, another form of knowing, which he calls *zhi xing* (知性, 'cognitive mind'), is indirectly suppressed—and this, according to his reading, is the reason why logic, mathematics, and science were not well developed in China.

The accuracy of Mou's classification is debatable, although once we understand the Kantian background, and appreciate the underlying mission that Mou had set himself, it may seem reasonable. Mou wanted to show that it is possible to develop the 'cognitive mind' from what in traditional Chinese philosophy is called *Liangzhi* (良知), meaning conscience, or knowing of the good, and which involves a certain 'self-negation'.

---

256. Mou Zongsan, *Collected Works 5, Philosophy of History*, 205.

He believed that this focus on *Liangzhi* owed to the fact that, within the Chinese tradition, philosophy aims to experience a cosmological order which is far beyond any phenomenon. *Liangzhi* comes from Mencius, and was further developed by the great Neo-Confucian Wang Yangming (王陽明, 1472–1529). In Wang's version we find a metaphysics that is much richer than Mencius's, which limited itself to the moral implication of *Liangzhi*. For Wang, *Liangzhi* is not knowing, but knowing everything (無知而無不知), and is furthermore not limited to the human being but also applies to other beings in the world such as plants and stones (草木瓦石 也有良知). That is not to say that *Liangzhi* exists everywhere, but that one can *project Liangzhi* into every being:

> When I say *zhi zhi ge wu* [致知格物, to study the phenomena of nature in order to know the principles], it means directing the *liangzhi* everywhere. The Liangzhi of my heart is the reason of Heaven [*tian li*]. By directing the *tian li* of *liangzhi* into things, they also acquire the reason. Directing the *liangzi* of my heart is *zhi zhi* [to know]; everything that acquires reason is *ge wu* [格物, contemplating the thing]. Therefore *xin* [heart] and *li* [reason] are combined.[257]

The supreme level of knowing consists in the *conscious return* to the *liangzhi* (良知) and its *projection* into every being (格物). *Liangzhi*, in this interpretation, becomes the *cosmic*

---

257. Wang Yangming, 'A Reply to Gu Dong Qiao' (答顧東橋書), in *Collected Works*, vol. 2 (王陽明集, 卷二) (Shanghai: Shanghai Ancient Works Publishing, 1992), '若鄙人所謂致知格物者, 致吾心之良知於事事物物也。吾心之良知, 即所謂天理也。致吾心良知之天理於事事物物, 則事事物物皆得其理矣。致吾心之良知者, 致知也。事事物物皆得其理者, 格物也。是合心與理而為一者也'.

mind, which has its origin in Confucius's teaching of *ren* (仁, 'benevolence'). The cosmic mind is an infinite mind. Here Mou combines Buddhism with Wang's thought and achieves a certain coherence of thought, or what is called *tong* (統, integration in a systematic sense). The question is as follows, then: If what occupies itself with *liangzhi* is a moral subject rather than a knowing subject, and if objective knowing has no position in *liangzhi*, then does this explain why there was no modern science and technology in China, allowing us then to conclude that, if China continues to rely on its classical Confucian teaching, it will never be able to develop any science and technology? This is the dilemma of New Confucianism: how to affirm Confucian teaching and at the same time allow modernisation to proceed, while not presenting the two as a separated *tong*. The response we will examine here consists in taking the most sophisticated part of the thought of Mou Zongsan, while admitting that it nonetheless harbours certain weaknesses and hence compromises his project of modernisation.

§18.2. THE SELF-NEGATION OF LIANGZHI IN MOU ZONGSAN

Mou further developed the concept of the *self-negation or self-restriction of Liangzhi* (良知的自我坎陷) found in the *I Ching* and in Wang's Neo-Confucianism. Here we follow Jason Clower's English translation of the term *Kanxian* as 'self-negation',[258] although it is not very exact. *Kanxian* is also a fall, like Heidegger's *Verfallen*; however Mou uses a very active mode here, suggesting a kind of selfhood (自我).

---

258. Mou Zongsan, *Late Works of Mou Zongsan. Selected Essays on Chinese Philosophy*, tr. J. Clower (San Diego, CA: California State University Press, 2014).

It is not simply given; rather, it demands a 'conscious falling'. Hence this falling is not a fault, but rather the realisation of the possibility of *Liangzhi*. One can perhaps discern a sort of Hegelian dialectics here, but this movement of thought can also be read in terms of a Kantian aesthetic judgement, in the sense that it is a heuristics—this is, however, not very clear in Mou's own writing. Sometimes he calls this action *zhi* (執), a word used often in Buddhism to describe the will to hold something instead of letting it go, or simply attachment. In this respect it has less to do with negation in the Hegelian sense and is more of a voluntary holding. Sticking to Kantian language, we might say that, for Mou, the relation between *Liangzhi* and what falls outside of it is not constitutive, but regulative. *Liangzhi* constantly negates and restricts itself, in order to arrive at its destination through a necessary detour:

> Therefore, self-negation, in order to become the subject of cognition, must be the conscious determination of the moral subject. This detour is necessary, since only by detouring thus can it reach its goal. Hence we call it 'to reach by detour [曲達]'. This necessity is the necessity of dialectics; this detour is the detour of dialectics, not merely the linear trajectory of intellectual intuition or a sudden awakening.[259]

The notion of 'achieving' or 'realisation' (達) is associated with the Neo-Confucian idea that there is a linear and direct relation between *Liangzhi* and knowledge. However, it is also clear that *Liangzhi* did not give rise to the form of knowledge that we call science. With this concept of the *self-negation of*

259. '故其自我坎陷以成認知的主體（知性）乃其道德心願之所自覺地要求的。這一步曲折是必要的。經過這一曲，它始能達，此之謂「曲達」。這種必要是辯證的必要，這種曲達是辯證的曲達，而不只是明覺感應之直線的或頓悟的達，圓而神的達'. Mou Zongsan, *Collected Works 21*, 127.

Liangzhi, Mou is able to announce that the knowing subject is only one possibility of *Liangzhi*, and hence that it is possible to have two minds at the same time. Here Mou uses a Buddhist expression, *One mind opens two gates* or *one mind two aspects* (一心開兩門),[260] meaning that the cosmic mind is able to negate itself in order to be a cognitive mind—an act of negation that would enable it to develop science or technology. The phenomenon belongs to the knowing mind, the noumenon to the cosmic mind, which is also the source of what Kant calls intellectual intuition; and yet

> it cannot really hold itself or persist; since when it holds, it is no longer itself, but the light of the intellectual intuition is hindered and turns aside, therefore it is its own shadow but not itself—that it is to say, it becomes the 'subject of cognition'. Therefore, the subject of cognition is what appears when light is hindered, and projects in a different way, and consequently the light of the intuition becomes other cognitive activities, which are analytic activities. Sensibility and cognition are only two modes of a knowing subject; the subject of cognitive knowing is the self-negation of the subject of intellectual intuition.[261]

Mou believes that with this concept of self-negation it is possible to systematically integrate Western philosophy—in a Kantian sense, the theory of knowledge—into the Chinese

---

260. This phrase comes from the Buddhist classic *Awakening of Faith in the Mahayana* (大乘起信論).

261. '但它並不能真執持其自己；它一執持，即不是它自己，乃是它的明覺之光之凝滯而偏限於一邊，因此，乃是它自身之影子，而不是它自己，也就是說，它轉成「認知主體」。故認知主體就是它自己之光經由一停滯，而投央過來而成者，它的明覺之光**轉**成為認知的了別活動，即思解活動。感性與知性只是一認知心之兩態，而認知心則是由知體明覺之自覺地自我坎陷而成者，此則等於知性'. Mou Zongsan, *Collected Works 21*, 127–8.

noumenal ontology. In doing so, Mou proposes several further 'translations' that may seem odd to Western philosophers. Firstly, he identifies the noumenon with the ontological, and the phenomenon with the ontic, in the Heideggerian senses of these terms (Mou had read Heidegger's *Kant and the Problem of Metaphysics* [1929], and hence integrates Heidegger's vocabulary into his division of systems). Secondly, he equates theological transcendence in Kant's philosophy with the Heaven of classical Confucianism. In so doing, Mou develops a very clean division of systems between the East and the West, but at the same time integrates the West into the possibilities of the East.

What is also important in Mou's analysis of the *Liangzhi* is a return to the political philosophy of Confucianism, namely *neisheng waiwang* (內聖外王, 'inner sageliness–outer kingliness'). This Confucian schema follows a linear trajectory that we met earlier: investigation of things (格物), extension of knowledge (致知), sincere in thoughts (誠意), rectify the heart (正心), cultivate the people (修身), regulate their families (齊家), govern well the States (治國), world peace (平天下). But the New Confucians understood that there was a problem with this direct projection from the inner to the outer. If, in the past, one trusted in a linear progression from the Emperor's cultivation of virtue and morality to the achievement of a peaceful world, this is no longer possible; instead, the projection now demands a detour through the outside. In other words, the traditional way of projection is no longer a progression but rather a regression. Hence a different trajectory is needed, and this resonates with the detour that *Liangzhi* has to take. This is very clear in Mou's book on political philosophy, *Dao of Politics and Dao of Governance* (政道與治道) (1974), in which he writes:

> Outer kingliness is the outward movement of inner sageliness,
> that is right. But there are two ways of achieving it, directly or
> via a detour. The direct approach is what we spoke of in the older
> time, the indirect approach (detour) is what we speak of now
> in relation to science and democracy. We think that the indirect
> approach allows the outer kingliness to be most expressive. But
> in the case of the direct approach, it becomes a shrinking back.
> Therefore, from inner sageliness to outer kingliness, when it is
> indirectly achieved, then there is a radical transformation, which
> doesn't come from direct reasoning.[262]

What Mou is suggesting here is that the ancient schema can
no longer function, and that therefore any enterprise that
seeks to start again with the ancient texts and with personal
cultivation (although these are still important) is no longer
sufficient. In comparison to the traditional conception of the
relation between politics and the moral, he perceives that
one must rethink this passage by affording a higher priority
to science and technology—in other words, he implicitly suggests that the 'detour' must in fact lead through the *Qi* or the
externalisation of the spirit.

Mou's philosophical task in relation to the question of technology ends here. Unlike others, he brings it into a metaphysical
register which is compatible with the Kantian system as well as
the traditional Chinese philosophy. Yet he goes no further, since
at bottom his thought is an idealist gesture. Mou insisted that

262. '外王是由內聖通出去，這不錯。但通有直通與曲通。直通是以前的
講法，曲通是我們現在關聯著科學與民主政治的講法。我們以為曲通能盡
外王之極致。如只是直通，則只成外王之退縮。如是，從內聖到外王，在
曲通之下，其中有一種轉折上的突變，而不是直接推理。這即表示：從理
性之運用表現直接推不出架構來表現'. Mou Zongsan, *Collected Works 9* (《
政道與治道》) (Taipei: Students Books Company, 1991), 56; cited by Zheng
Jiadong (鄭家棟), *Mou Zongsan* (《牟宗三》) (Taipei: Dongda Books, 1978), 81.

Kant's philosophy is by no means a transcendental idealism, but rather an empirical realism; and, like the Neo-Confucians, he held that mind and things cannot be separated. Yet in Mou's work, the mind becomes the ultimate possibility of knowing both phenomenon and noumenon. What conditions the mind to be such a pure starting point? Like Fichte and Schelling, Mou identifies *Liangzhi* as the unconditioned, with the fundamental difference that *Liangzhi* is not a cognitive *Ich*, but rather a cosmic *Ich*. If *Liangzhi* can negate itself into a knowing subject, then the knowing subject, thus derived from a conscious act of *Liangzhi*, dwells in a coherent relation to *Liangzhi*. Hence when science and technology are developed in this way, they will be *a priori ethical*. To put it in another way, in relation to the *Qi-Dao* discourse, we might say that *Qi* is a *possibility* of *Dao*. Hence the relation between *Qi* and *Dao* is not one of 'use', but is instead a recursive relation. This is also the reason I consider Mou's approach still to be an idealist one.

So how useful is Mou's strategy in reconsidering the modernisation project? Mou's biographer Zheng Jiadong noted that

> For hundreds of years, maintaining the status quo of the nation and at the same time being able to absorb Western knowledge— having both fish and the bear's paw—was what the Chinese dreamt of. The 'negation of *Liangzhi*' is the most sophisticated and philosophical expression of this dream. But whether this dream can be realised is another question.[263]

And indeed, Mou's 'idealist' proposal for such a metaphysical and cultural transformation was totally ignored by the materialist movement in mainland China, a movement he had heavily

263. Zheng Jiadong, *Mou Zongsan*, 89.

criticised. However, it is lamentable that Mou's philosophical project wasn't taken further. In mainland China, Mou's work has not been well received, owing to his critical view of Communism: for him, it had very little to do with the Chinese tradition and indeed, on the contrary, only succeeded in destroying that tradition. Instead, another path was followed, in the name of the dialectics of nature, which leads to what I call the end of *Xing er Shang Xue* (the ancient expression that is used to translate the English word 'metaphysics'), and the emergence of a new discipline of Science and Technology Studies.

§19. THE DIALECTICS OF NATURE AND THE END OF XING ER SHANG XUE

Martin Heidegger declared the end of metaphysics on various occasions: he considered Nietzsche to have been the last metaphysician. In his 1969 essay 'The End of Philosophy and the Task of Thinking' he declared that the end of philosophy was portended by the beginning of cybernetics. This 'end', however, is not universal, although as we will see, it is a general tendency brought about by modern technology—an end that I characterise as 'dis-orientation'. The 'end of metaphysics' did not take place simultaneously in the West and in the East: firstly because 'metaphysics' is not equivalent to its usual translation in Chinese, *Xing er Shang Xue*—as we have seen clearly above, the development of *Xing er Shang Xue* was not able to produce modern science and technology; and secondly because, in the East, the end of *Xing er Shang Xue* took another form: the disassociation of *Dao* from *Qi*. For China, this end has only become present as a kind of aftershock over the past century, as if it had been deferred, and only arrived when a new destiny was imposed—modernisation, and later globalisation—a process in which Chinese philosophy no longer plays any important role—or only in the promotion of tourism and the culture industry.

'Needham's question' continued to haunt Chinese scholars throughout the twentieth century. If one follows the logic of Needham and Feng, one might say that there was never a *philosophy* of technology in China before the twentieth century. As we have seen, in one sense, China has only a philosophy of nature along with a moral philosophy, which may regulate how technical knowledge is acquired and applied. In Europe, it could be argued, philosophy of technology was only inaugurated in the late nineteenth century, and initially gained its place in academic philosophy in Germany through the works of Ernst Kapp, Martin Heidegger, Friedrich Dessauer, Manfred Schröter and others. However, as we saw above, the question of technics has always been present in Western philosophy, and indeed could be said to be cosmotechnically constitutive of Occidental thinking—even if it is in some sense repressed, if we follow Bernard Stiegler's argument, which we will discuss in detail in Part 2 below.

In China a different trajectory was followed, mainly owing to the fact that, from 1949 on, Marxist ideology became dominant in all aspects of the new Republic. Engels's *Dialectics of Nature*, together with his *Anti-Dühring*, were widely studied and were presented as the foundational theory for the development of socialist science. From the time of its translation into Chinese in 1935, *Dialectics of Nature* became a 'discipline' in China, equivalent to Science and Technology Studies in the West.[264] In these two books, Engels seeks to show that a materialist dialectics should become the main method for natural science.

---

264. Lin Dehong (林德宏), *Fifteen Lectures on Philosophy of Technology* (科技哲學十五講) (Beijing: Peking University Press, 2014).

Anti-Dühring was also a response to the 'degeneration of Berlin Hegelianism', where the idealist and metaphysical interpretation of nature had become predominant. In the second preface to *Anti-Dühring* Engels writes:

> Marx and I were pretty well the only people to rescue conscious dialectics from German idealist philosophy and apply it in the materialist conception of nature and history. But a knowledge of mathematics and natural science is essential to a conception of nature which is dialectical and at the same time materialist.[265]

Engels's materialist dialectics starts from empirical facts and sees nature as a constant process of evolution. We might simplify it into two main points. Firstly, Engels wants to argue that every natural being has its history, from plants to animals to nebulae. Engels praised Kant's *Universal Natural History and Theory of the Heavens* (1755), in which Kant had already suggested that the formation of the earth and the solar system was an evolutionary process. If this is the case, then according to the Kantian cosmology, all beings on earth and in the universe must also come into being in time. As Engels wrote, 'Kant's discovery contained the point of departure for all further progress'.[266] Secondly, in the spirit of Marx, Engels wants to show that there is a 'humanised nature', a nature perceived by the human being through his labour. The second point had significant influence in China, probably because the chapter 'The Part Played by Labour in the Transition from Ape to Man', which elaborates on Darwin's evolutionary theory,

---

265. F. Engels, in *Marx and Engels Collected Works* (London: Lawrence & Wishart, 50 vols, 1975–2004), vol. 25, 11.

266. Ibid., 324.

was separately translated and appeared before the whole manuscript had been published in Chinese. In this chapter, Engels emphasises that animals don't have tools, and hence they can only *use* nature, while humans, after the liberation of the hands, are able to use tools and hence to *master* nature. The Marxist philosopher and economist Yu Guangyuan (於光遠, 1915–2013), widely known as the figure who most profoundly influenced Deng Xiaoping's economic reform, led the translation of Engels's *Dialectics of Nature*, and also in his own work extended this 'humanised nature' into a more concrete concept of 'social nature' as a second nature, and also as a new 'discipline'.[267]

During the Civil Wars in China (1927–1937, 1945–1950), and later owing to the deterioration of relations between the People's Republic of China and the Soviet Union, China was forced to develop science and technology from the fragmented and insufficient knowledge that the country had at the time. In 1956, Yu Guangyuan, together with some natural scientists, drafted a 'Twelve Year (1956–1967) Research Plan of Dialectics of Nature (Philosophical Questions in Mathematics and Natural Science)', and in the same year established a regular newsletter. Engels's *Dialectics of Nature* became the guiding method of a national movement proposed by Mao in 1958: 'open fire against nature, carry out technological innovation and technological revolution'.[268] At this point, then, *Dialectics of Nature* became not just a critique of the 'degeneration of Hegelianism' and the

---

267. Yu notably published a book entitled *A New Philosophical School is Emerging in China* (一个哲學學派正在中國興起) (Nanchang: Jiangxi Science and Technology Publishing House, 1996).

268. '向自然界开火，进行技术革新和技术革命. '

'abuse of science' in Germany, but also a method by which to understand nature and hence to 'master' it.

The Cultural Revolution (1966–1976) on the one hand further destroyed the traditions that were regarded by the regime as 'regression' according to Marx's theory of the progress of history (Primitive Communism–Slavery–Feudalism–Capitalism–Socialism–Communism); and on the other hand made *Dialectics of Nature* the foundation of science and technology in China. In 1981 the Chinese Society for Dialectics of Nature (CSDN) was established under the approval of Deng Xiao Ping. The influence of *Dialectics of Nature* was hence extended beyond science to technology studies, becoming a 'weapon' to improve productivity in all domains. The philosopher Chen Changshu (陳昌曙, 1932–2011) may be said to be responsible for formally and officially founding the discipline of 'Philosophy of Science and Technology' in China. He proposed to the Academic Degrees Committee of the State Council in 1990 the adoption of this name for the discipline, in place of 'dialectics of nature'.[269] Chen's own *Introduction to the Philosophy of Technology* (1999) is a valuable textbook in the field.[270] But although this newly formalised discipline took on a new name, *Dialectics of Nature* was still its historical foundation stone, despite the fact that, apart from the chapter on evolution, Engels's book contains *nothing* about technology.

Philosophy of Science and Technology is hence rather new in China, but it has a strong dynamic behind it, owing to the recognition of the importance of this subject. For example,

269. Xia Li, 'Philosophy of Science and STS in China: From Coexistence to Separation', *East Asian Science, Technology and Society: An International Journal* 5 (2011), 57–66.

270. Chen Changshu (陳昌曙), *Introduction to Philosophy of Technology* (技術哲學導論) (Beijing: Science Publishing, 1999).

among others, the philosopher Qiao Ruijin's *An Outline of Marxist Philosophy of Technology* (2002) systematically explores the appropriation of Marxist critique of technology in China; and Lin Dehong's *Man and Machine: The Essence of High Technology and the Renaissance of the Humanities* elaborates on the possibility of a new humanities which takes technology into account.[271] Although I am sympathetic to these efforts, it strikes me that there has been a lack of continuity or even coherence in thinking China and its relation to technics. That is to say that, apart from the recent work of Li Sanhu, such a philosophy of technology has only ever been an attempt to introduce *Technikphilosophie* or Philosophy of Technology into China, in parallel with the Marxist critique of technology, in order to assimilate them. This is the case with the names we have cited above, and other contemporaries such as Carl Mitcham, Herbert Marcuse, Andrew Feenberg, Albert Borgmann, and Hubert Dreyfus—*as if China and Europe have the same understanding of technics*. The universalisation of European philosophy is thus *pharmacological* in the sense that, although it may lead to broader conversations, its dominance can also close down any path to a more profound dialogue.

This, then, is what we may call the end of metaphysics as *xing er shang xue*: the metaphysical thinking that, in Chinese thought, maintains the coherence of the human-cosmological system is interrupted in such a way that a metastability can no longer be restored. I call this situation 'dis-orientation' in two senses: firstly, there is a general loss of direction: one finds oneself in the middle of the ocean without being able to see

271. Lin Dehong (林德宏), *Human and Machine: The Essence of High Technology and the Renaissance of the Humanities* (人与机器—高科技的本质与人文精神的复兴) (Nanjing: Jianshu Education Publishing, 1999).

either point of departure or destination—the scenario that Nietzsche depicted in *The Gay Science*; secondly, unlike the Occident, the Orient is negated in such a way that it ceases to be the Orient, and in consequence the Occident also loses sight of the Orient. In other words, a homogeneity is brought about by technological convergence and synchronisation. Philosophies of technology in China over the past thirty years have been active responses to technological globalisation and economic growth in China, but the tendency to identify the Chinese concept of technics with that of the West, or to allow the latter to override the former, is a *symptom* of globalisation and modernisation, one that amplifies the tendency of forgetting and the detachment from the question of cosmotechnics—a question which, therefore, in China, has been subject to its own 'forgetting' which is not the same as that described by Heidegger.

Technological reason is expanding to the extent that it is becoming the condition of all conditions, the principle of all principles. A totality is in the process of forming through technical systems, as Jacques Ellul already predicted in the 1970s.[272] If this technological reason is to be resisted, this can only be done by bringing forth other forms of reasoning to constitute a new dynamics and new order. Accelerationism appeals to a universalism that it attempts to decouple from any colonialist imposition of culture. Yet at the same time it draws this universalism from a 'Promethean' conception of

---

272. J. Ellul, *The Technological System*, tr. J. Neugroschel (London: Continuum, 1980); this book can be read as an extension of Simondon's *Du Mode d'existence des objets techniques*. For further analysis see Y. Hui, 'Technological System and the Problem of Desymbolization', in H. Jerónimo, J. L. Garcia, and C. Mitcham (eds), *Jacques Ellul and the Technological Society in the 21st Century* (Dordrecht: Springer, 2013), 73–82.

technology which it champions but whose cultural specificity it never subjects to interrogation. Here the category of technics itself is exhaustive, and harbours only one destiny. Beyond such an accelerative universalization, the diversity of technicities and their various relations to nature—as well as to the cosmos—has to be rediscovered and reinvented. The only hope for China to avoid the total destruction of its civilisation in the Anthropocene is to invent a new form of *thinking* and *invention*, as Mou Zongsan did, but this time in a different way. This will require it to distance itself from the traditional idealist approach and to look for another interface between what Mou called noumenal and phenomenal ontology. To achieve this requires thinking *cosmotechnically*, and developing a form of thinking that allows a further development of *Qi* without detaching it from *Dao* and cosmological consciousness. In Part 2, we will take up this question through a reinterpretation of time and modernity.

PART 2:
MODERNITY AND
TECHNOLOGICAL CONSCIOUSNESS

§20. GEOMETRY AND TIME

In Part 1 we demonstrated that, even if what Western thought would recognise as a 'philosophy of technology' remained alien to the Chinese, nevertheless the exposition of the history of the relation between *Qi* and *Dao* enables us to unearth a 'technological thinking' in Chinese philosophy. It is our task in Part 2 to ask what happened when this Chinese technological thinking confronted the Western one, grounded in its long philosophical tradition. What is called 'modernity' in Europe didn't exist in China, and modernisation only occurred after the confrontation between the two modes of technological thought. Here this confrontation will be described as the tension between two temporal structures; but this will also involve a rethinking of the question of modernity itself. Over the course of the twentieth century, the voices that proclaimed the necessity of 'overcoming modernity' were echoed firstly in Europe, and then in Japan—though with different motives—and now are heard almost everywhere, in light of ecological crisis and in the wake of technological catastrophes. But what these voices ended up calling forth—as seems to be forgotten among anthropologists who propose a return to ancient cosmologies or indigenous ontologies—was war and metaphysical fascism. It is by revaluating the question of modernity through the confrontation of the two modes of thought mentioned above that I want to suggest that it is not at all sufficient to go back to 'traditional ontologies', but that we must instead reinvent a cosmotechnics for our time.

Had Needham already answered the question of why modern science and technology did not arise in China? Did Chinese intellectuals answer Needham's question in a satisfactory way in the twentieth century? Needham certainly provided a very systematic analysis of different factors, amounting to far more than mere social constructivism. His analysis took in the system of public recruitment of government officials, philosophical and theological factors, and social-economic factors, all of which had significant impact on the formation of a singular culture. These factors form an assemblage which expresses the tendencies, forces, and contingencies that constitute Chinese history. Yet I fear that Needham's analyses are not sufficient to explain the lack of modern science and technology, and that there is something more fundamental at stake in the Chinese philosophical system; and in order to apprehend this, we will have go deeper. As we have seen, Chinese philosophy is based on an organic rather than a mechanical form of thinking—something Needham emphasised, but pursued no further. Mou Zongsan, in turn, suggested that Chinese philosophy is characterized by a focus on noumenal ontology, as indicated by the tendency to turn experience toward the infinite. It seems that in the Chinese philosophical mentality, the cosmos has a rather different structure and nature than in that of the West; and that the role of the human and its way of knowing are also determined in a different way, in coherence with the cosmos.

As we shall see below, according to the observations of Sinologists, the ancient Chinese did not develop a systematic geometry—the knowledge of space[1]—and neither did they elaborate on the theme of time. Below we will explore the implications of the thesis that Chinese thinking is marked by an absence of any axiomatic system of geometry and an under-elaboration of time.

---

1.   B. Stiegler and E. During, *Philosopher par accident* (Paris: Galilée, 2004), 52.

§20.1. THE ABSENCE OF GEOMETRY IN ANCIENT CHINA

Needham noted that in ancient China there was no geometry, but only algebra.[2] Of course, this is not to say that there was no geometrical knowledge—indeed there was, since the history of China can also be read as the history of controlling two rivers (the Yangzhi River and Yellow River) prone to constant flooding and occasional drought. Managing these two rivers must necessarily have demanded geometrical knowledge, measurement, and calculation. Rather, Needham means that a systematic knowledge of geometry came rather late, possibly not until after the translation of Euclid's *Elements of Geometry* by the Jesuits towards the end of the seventeenth century. Some historians suggest that the *Jiu Zhang Suan Shu* (九章算術, *Night Chapters on the Mathematical Art*, 10th–2nd centuries BC) and the commentary of the Mathematician Liu Hui (劉徽, 3rd Century) already demonstrated an advanced geometrical thinking.[3] However, the latter differed fundamentally from Greek geometry in the sense that the *Jiu Zhang Suan Shu* did not establish a formal deductive system of axioms, theorems, and proofs; and in fact, 'unlike ancient Greek mathematics, which emphasises geometry, the achievement of ancient Chinese mathematics lay primarily in calculation'.[4] Other historians have shown that what is lacking in ancient Chinese mathematics is the development of a

---

2.   Needham, 'Poverties and Triumphs of the Chinese Scientific Tradition', 21.

3.   Mei Rongzhao, 'Liu Hui's Theories of Mathematics', in Fan Dainian and R.S. Cohen (eds), *Chinese Studies In the History and Philosophy of Science and Technology* (Dordrecht: Springer, 1996), 243–54: 248.

4.   Ibid., 244.

'complete structural theoretical system'.[5] For example, Zhang Heng (78–139) is considered to have postulated that the sun, moon, and planets move along spherical paths, but owing to the lack of any axiomatic system, the discovery was not developed any further. Geometry and logical systems only started to emerge in China during the seventeenth century, following the translation of Euclid's Geometry (*Jihe Yuanben*) by Matteo Ricci and Paul Xu Guangqi. Xu Guangqi perceived that 'logic is a forerunner of other studies and a prerequisite for the understanding of various other disciplines', and therefore endeavoured to make geometry and logic the cornerstone of the new science.[6]

Of course, geometry was a significant discipline in ancient Greece, and the philosophical rationalisations of the Ionian philosophers were closely related to its invention. Thales, the first known Ionian philosopher and pioneer of geometry, used his knowledge of the geometrical properties of triangles to calculate the height of pyramids and to determine the diameters of the sun and the moon. Thales's assumption that the world is composed of a homogeneous element is a necessary precursor to the geometrical investigation of order, measure, and proportions.[7] And we should not forget that, at least according to Hippolytus, Pythagoras united astronomy, music,

5.   Jin Guantao, Fan Hongye, and Liu Qingfeng, 'The Structure of Science and Technology in History: On the Factors Delaying the Development of Science and Technology in China in Comparison with the West since the 17th Century (Part One)', in Dainian and Cohen (eds), *Chinese Studies in the History and Philosophy of Science and Technology*, 137–164: 156.

6.   Jin Guantao, Fan Hongye, and Liu Qingfeng, 'Historical Changes in the Structure of Science and Technology (Part Two, a Commentary)', in ibid., 165–83.

7.   P. Clavier, 'Univers', in D. Kambouchner (ed.), *Notions de Philosophie*, I (Paris: Gallimard, 1995), 45.

and geometry.[8] This rationalisation is also central to the cosmogony in Plato's *Timaeus*, in which god becomes a technician who works on the receptacle (*chōra*) according to different geometrical proportions. It was this spirit that led to the great achievements of Greek geometry. Such rationalization reached its height in the system laid down by Euclid of Alexandria, in which a mathematical discipline is described as a collection of axioms, and where the theorems derived from them can be ascertained to constitute a complete and coherent system.

Notwithstanding their advances in geometry, it has often been noted that the ancient Greeks were not so strong in algebra. One of the best demonstrations of this is Archimedes's *On Spirals*, in which the mathematician mechanically describes how to trace a spiral without employing any symbol or equations. As mathematician John Tabak observes, the 'Greeks had little interest in algebra. Our facility in generating new curves is due largely to our facility with algebra'. By the time of Pappus of Alexandria, the last of the great ancient Greek geometers, they had already achieved a quite comprehensive understanding of lines, planes, and solids, yet '[f]or the Greeks describing almost any curve was a struggle'.[9] During the Middle Ages, research in geometry slowed as it fused with theology, although geometry was still regarded as one of the seven liberal arts. What is significant during this period is the reintroduction of Greek geometry to the Romans, as indicated firstly by the translation of Euclid's *Elements* from Arabic into Latin by Adelard of Bath (1080–1152) around 1120, and later, the

---

8.   C. Riedweg, *Pythagoras, His Life, Teaching, and Influence* (Ithaca, NY and London: Cornell University Press, 2002), 25.

9.   J. Tabak, *Geometry: The Language of Space and Form* (New York: Facts on File, 2004), 36.

first translation from Greek to Latin by Bartolomeo Zamberti (1473–1543) at the end of the fifteenth century.[10] During the Renaissance, geometry was partly driven by artistic creation, especially painting: the techniques developed to project a three-dimensional object onto a two dimensional plane, and the theory of perspective, led to what we know today as projective geometry. In the sixteenth and seventeenth centuries, the rise of modern science in Europe, as exemplified by the work of Kepler, Galileo, and Newton, can be characterised as a spirit of geometrisation. In a 1953 remark that has often been quoted, by Needham among many others, Albert Einstein observed that

> The development of Western science has been based on two great achievements: the invention of the formal logical system (in Euclidean geometry) by the Greek philosophers, and the discovery of the possibility of finding out causal relationships by systematic experiment (at the Renaissance). In my opinion one need not be astonished that the Chinese sages did not make these steps. The astonishing thing is that these discoveries were made at all.[11]

Einstein's characterisation of geometry as a 'formal logical system' may remind us of our discussion of the development of Chinese thought in Part 1: as we saw, the school of Moism, which advocated logic and technics, was repressed by Confucians such as Mencius in favour of an outlook based on a moral

---

10. C. Scriba and P. Schreiber, *5000 Years of Geometry: Mathematics in History and Culture*, tr. J. Schreiber (Basel: Springer, 2015), 231, 236.

11. A. Einstein, Letter to J.S. Switzer, April 23, 1953, in A.C. Crombie (ed.), *Scientific Change: Historical Studies in the Intellectual, Social, and Technical Conditions for Scientific Discovery and Technical Invention, from Antiquity to the Present* (London: Heinemann, 1963), 142.

cosmology. The second achievement of the West, according to Einstein, was the discovery of causal relations through experimentation. This search for causal regularities and 'laws of nature' is a very specific form of philosophising about nature, one that moves from concrete experiences to abstract models. In relation to Chinese thought, Needham posed a very relevant question here: Can this emergence of the concept of laws of nature in Europe in the sixteenth and seventeenth century be attributed specifically to scientific and technological developments?[12] Catherine Chevalley answers in the affirmative by pointing out three key scientific developments in Europe during this period: (1) the geometrisation of vision (Kepler); (2) the geometrisation of movement (Galileo); and (3) the codification of the conditions of the experiment (Boyle, Newton). In each of these cases, geometry plays a crucial role in so far as it allows for a detachment of scientific knowledge from everyday experience. In the first instance, Kepler mobilised the Plotinian understanding of light as emanation against Aristotle's substantialist definition, and showed that the formation of images on the retina involves a complicated process which follows geometrical rules (i.e. diffraction and the geometrical deformation of inverted images). Similarly, Galileo's geometrisation of the laws of movement, which superseded the Aristotelian concept of change (*metabolē*) as modification of substance and accidents (generation or corruption), proceeded by considering an ideal environment of the void, where falling objects of different masses will acquire the same speed, against the intuitive belief

12.   J. Needham, 'Human Laws and Laws of Nature in China and the West I', *Journal of the History of Ideas* 12:1 (January 1951), 3–30 ; 'Human Laws and Laws of Nature in China and the West II: Chinese Civilization and the Laws of Nature', *Journal of the History of Ideas* 12:2 (April 1951), 194–230.

that an object with larger mass will fall at a higher speed.[13] The apodictic nature of geometry stands against the fallibility of intuition—a passage in Galileo's *Dialogue Concerning the Two Chief World Systems* reveals the striving for a methodological certitude that is not affected by the vicissitudes of human error and judgment:

> If this point of which we dispute were some point of law, or other part of the studies called the humanities, wherein there is neither truth nor falsehood, we might give sufficient credit to the acuteness of wit, readiness of answers, and the greater accomplishment of writers, and hope that he who is most proficient in these will make his reason more probable and plausible. But the conclusions of natural science are true and necessary, and the judgment of man has nothing to do with them.[14]

Einstein was not unjustified, then, in his assessment of the advance of geometry in Europe. In fact, if we look at the history of cosmology from its mythical origins up to modern astronomy, via Claudius Ptolemy, Copernicus, Tycho Brahe, Kepler, and Newton, at every stage it is fundamentally a geometrical question.[15] Even Einstein's theory of general relativity, which identifies gravity with the curvature of four-dimensional

---

13.   C. Chevalley, 'Nature et loi dans la philosophie moderne', in *Notions de Philosophie*, I, 127–230.

14.   Cited by C. Bambach, *Heidegger, Dilthey and the Crisis of Historicism* (Ithaca and London: Cornell University Press, 1995), 50.

15.   See H.S. Kragh, *Conceptions of Cosmos: From Myths to the Accelerating Universe: A History of Cosmology* (Oxford: Oxford University Press, 2013), in which Kragh formulates a history of the cosmos according to a passage from to Euclidean geometry to non-Euclidean geometry, e.g. Riemannian geometry.

space-time, is fundamentally a geometrical theory (albeit no longer a Euclidean one).

§20.2. GEOMETRISATION AND TEMPORALISATION

But rather than limiting ourselves to geometry as a mathematical subject, let us take the question further by connecting it with the question of time. It seems to me that the relation between time and geometry/space is fundamental to the Western concept of technics and its further development into efficient mnemotechnical systems. In posing the question in this way, we will shift from abstraction to idealisation—that is, from mental abstraction to idealisation in externalised geometrical forms. Idealisation has to be distinguished from ideation, which still concerns theoretical abstraction in thought—for example, we can think of a triangle (e.g. ideation), but the apodictic nature of the triangle becomes common to all when it is externalised (e.g. drawn).[16] Idealisation in this sense thus involves an exteriorization, whether through writing or drawing. My reasoning on the relation between geometry, time, and technics can be summarised as follows: (1) geometry demands and allows the spatialization of time, which involves (2) exteriorization and idealization through technical means, (3) geometrical apodicticity allows logical inferences as well as the mechanization of causal relations, and (4) the technical objects and technical systems made possible on the basis of such mechanisation in turn participate in the constitution of temporality: experience, history, historicity.

---

16.   This speculation emerged out of many long discussions with Bernard Stiegler, and I take the distinction between ideation and idealization from him.

Geometrisation is a spatialisation of time in various senses. Firstly, it visually expresses the movement of time (either in linear form or in a cone section); secondly, it both spatialises and exteriorises time in such a way that time can be recollected in the future in an idealised form (we will come back to this point later when discussing the thought of Bernard Stiegler). My hypothesis—though delicate and speculative—is simply the following: Not only was geometry not developed in China; in addition, the question of time was not addressed in the same way as in the West; and it is these two considerations together that gave rise to a different concept of technics in China, or indeed the apparent absence of any thinking of technics. This argument may seem rather perplexing at first glance. In order to explain, I will firstly give an outline of the question of time in China, and then move on to the relation between time and geometry, before we arrive at a synthesis of them in relation to technics.

Sinologists such as Granet[17] and Jullien have addressed the question of time in Chinese thought, and both argue that there is no concept of linear time in China but only *shi*, which means 'occasions' or 'moments'. The Chinese traditionally manage their lives according to *sìshí* (四時), meaning the four seasons.[18]

17. M. Granet, *La pensée chinoise* (Paris: Albin Michel, 1968), 55–71.

18. This is debatable, however: according to Chinese historian Liu Wenying, the classification of four seasons only arrived toward the end of the Western Zhou period (1046–771 BC). Previously, a year was divided into spring and autumn. See Liu Wenying (劉文英), *Birth and Development of the Concepts of Time and Space in Ancient China* (中國古代時空觀念的產生和發展) (Shanghai: Shanghai Peoples' Press, 1980), 8; Moreover, this has to be further justified, since it may be argued that from the Shang dynasty (1600 BC–1046 BC), there has been a system for recording days and years known as 'Stems-and-Branches' (天干地支), which functions according to a sexagenary cycle; moreover, this recording system was integrated with the *I Ching* for fortune telling, which also demands calculation; however, when Granet and Jullien

Jullien also observes that this conception of time is closely related to the *Huainanzi* (discussed in Part 1 above) and its schematic definition of the relation between political and social conduct and seasonal change. As he notes, Chinese culture's understanding of time, where the movement of the seasons is taken as a first principle, is fundamentally different from that of the Aristotelian tradition, which is based on a conception of time as movement from one point to another, or from one form to another, involving quantity and distance.[19] From antiquity,

---

argue that the concept of time was not elaborated in China, they mean that, although one can find ways of recording dates and years, the perception and understanding of time remained closely attached to concrete events rather than abstract time. Equally, the Chinese were pioneers in clockmaking: Zhang Heng (78–139) succeeded in using water to rotate an armillary sphere, and the polymath Su Song (1020–1101) constructed one of the first clocks in the world, the 'Water-powered Armillary & Celestial Tower' (1088). Therefore not only did the mechanization and calculation of time (calendrical science) exist already in the Han dynasty, it was very advanced (see J. Needham, Ling Wang, D.J. de Sollar Price, *Heavenly Clockwork The Great Astronomical Clocks of Medieval China* [Cambridge: Cambridge University Press, 2008], 7. Su's machine was abandoned 1214 due to the difficulty of transportation during the move of the capital [with the new dynasty], and no one else could understand the documents drafted by him in order to rebuild it.) Indeed, it is undeniable that China had a leading position in many technological domains before the sixteenth century. However, the question we should reflect on here is that of whether the existence of calendarity implies a conceptual 'elaboration' of time? One does not necessarily follow from the other.

19.   In Aristotle's *Physics*, time is considered to be 'quantity of movement' defined by the before and after, we can find a clear definition of time in 220b5-12, in which see that time considered as 1)movement; 2) number; 3) between: 'Time is the same at all places simultaneously, but earlier and later times are not the same, because also the present [stage of a] movement is just one, whereas the past and future [stages] are different [sc. from each other]. And time is a number, not by which we number, but rather as a thing numbered, and this is always different when earlier or later; for the nows are different. [Similarly] the number of 100 horses and of 100 men is the same number, but those of which it is the number—the horses and the men—are different', cited by D. Bostock, *Space, Time, Matter, and Form: Essays on Aristotle's Physics* (Oxford: Oxford University Press, 2006), 141,

time has been considered to be inter-momentary—that is, it is thought in terms of movement between one point and another (we may want to call this a primary spatialisation *qua* geometrisation, in contrast to a second spatialisation in writing which we will discuss below). For the ancients, time is 'between' (*metaxu*); for the Stoics it is 'interval' (*diastêma*); and for Augustine, *sentimus intervalla temporum*.[20] But, as Jullien shows, this notion of time as interval only reached China in the nineteenth century, following the adoption of the Japanese translation of time as 'between-moments'—*jikan* in Japanese and *shíjiān* (時間) in Chinese.[21]

An alternative, more encompassing concept of time is found in the Chinese understanding of the cosmos/universe or *Yu Zou* (宇宙),[22] where *Yu* is space and *Zou* is time. *Zou* is etymologically related to the wheel of a wagon, from whose circular movement time takes its figurative metaphor.[23] *Sìshí* is likewise cyclical, and is divided into twenty-four solar terms (節氣) indicated by seasonal change. For example, the period around 5–6 March is called *jingzhe* (驚蟄), literally meaning 'the awakening of insects', indicating the end of hibernation. In the *I Ching*, time (*shí*) is also referred to in terms of occasions: for example, one speaks of 'observing *shí*' (察時)', 'understanding *shí*' (明時)', 'waiting for *shí*' (待時)', and so on.[24] *Shí* is also associated with *shì* (勢), which Jullien translates as 'propensity' (*propension*), and which can be understood,

---

20. Jullien, *Du Temps*, 74.

21. Ibid., 73.

22. Both 'cosmos' and 'universe' are translated into Chinese as *Yu Zhou*.

23. Liu Wenying, 21–2.

24. Chun-chieh Huang (黃俊傑), *Confucian Thought and Chinese Historical Thinking* (儒家思想與中國歷史思維) (Taipei: Taiwan University Press, 2014), 3.

simplifying somewhat, as situational thinking.[25] (Following the work of Marcel Detienne and Jean-Pierre Vernant, Jullien also pointed out that a similar thinking could be identified in ancient Greece, bearing the name *mētis*, which Detienne and Vernant gloss as 'cunning intelligence'.[26] Although the Sophists explored the concept of *mētis*, this mode of thought was repressed and excluded from 'Hellenic science'). The association between the two concepts *shí* and *shì*, for Jullien, also undermines the idealist tendency to think from the subject or *I,* tending rather toward what he calls a *transindividual* relation with the outer world: what constitutes the subject is not the will or the desire to know, but rather what is outside of it and traverses it. [27]

We may therefore wonder whether, whereas in Chinese thinking, truth did not constitute a veritable philosophical question, while the search for apodicticity among the Greek thinkers allowed geometry to become the primary mode of representation of the cosmos (time and space), and thus allowed the reconstitution of the temporalisation of experience by means of technics. Bernard Stiegler argues that the relation between geometry and time in the West is demonstrated in Socrates's response to Meno's question concerning virtue, where he shows that geometry is essentially technical and temporal in the sense that it demands a writing and a schematisation. Stiegler skilfully reconstructs the question of geometry as a question of time, or, we may say, a question

25.   F. Jullien, *Traité de l'efficacité* (Paris: Éditions Grasset, 1996). Granet also emphasised this point, describing the concept of space in China as 'rhythmic and geometric'; however, one should also bear in mind that he was not really talking about space, but rather *fengshui*.

26.   M. Detienne and J.P. Vernant, *Cunning Intelligence in Greek Culture and Society*, tr. J. Lloyd (Chicago: University of Chicago Press, 1991).

27.   Jullien, *Du Temps*, 84.

of re-temporalisation. Recall that in the *Meno*, Socrates is challenged by Meno with a paradox: if you already know what virtue is, then you don't need to look for it; however, if you do not know what it is, then even when you encounter it, you will not be able to recognise it. The conclusion that follows is that one can never know what virtue is. Socrates replies to this challenge with a ruse: he says that he once knew what virtue is, but has forgotten, and hence will need help to remember. Socrates demonstrates this process of remembering or *anamnesis* by asking a young uneducated slave to solve a geometrical problem by drawing it in the sand. For Stiegler, this operation exemplifies the technical exteriorisation of memory: it is only the markings on the sand—a form of *technē*—that allow the slave to trace the lines of the problem and to 'remember' the forgotten truth. As Stiegler notes, geometrical elements such as a point or a line do not really exist, if we understand existence in terms of spatial-temporal presence. When we draw a point or a line in the sand, it is no longer a point, since it is already a surface. The ideality of geometry demands a schematization qua exteriorization as writing:[28]

> Geometry is knowledge of space, and space is a form of intuition.
> Thinking of space as such an a priori form suppose this capacity
> of projection that the figure represents. But it is essential here to
> notice that this projection is an exteriorization not only in that it
> allows a projection for intuition, but also in the sense that it con-
> stitutes a retentional space, that is to say a support of memory

---

28. See Stiegler and During, *Philosopher par accident*, chapter 2.

which, step by step, backs the reasoning of the *temporal flux* that is reason, which thinks.[29]

According to Stiegler's deconstruction, then, the Platonic concept of truth as recollection is necessarily supplemented with a technical dimension which, however, Plato does not thematise. Stiegler calls this 'tracing of the line on the sand', this exteriorised memory, *tertiary retention*—a term that he adds to the primary and secondary retention explained in Husserl's *On the Phenomenology of the Consciousness of Internal Time*.[30] When we listen to a melody, what is retained immediately in memory is the primary retention; if tomorrow I recall the melody, this testifies to a secondary retention. What Stiegler calls tertiary retention, then, would be, for example, the musical score, the gramophone, or any other recording device that externalises the melody in a stable and enduring form outside of consciousness proper.

Here Stiegler takes up the thread of Jacques Derrida's *Introduction to Husserl's Origin of Geometry*, where Derrida confirms that what constitutes the origin of geometry is communication from generation to generation, as Husserl himself argues; but adds that this is only possible through writing, which assures the 'absolute traditionalisation of the object, of its absolute objectivity'. Geometry is constituted not only by communication (drawn figures), but is itself the constituent of communication (ortho-graphs), without which the 'self-evidence' or apodicticity of geometry would not be retained.[31]

---

29. Ibid., 52.

30. E. Husserl, *On the Phenomenology of the Consciousness of Internal Time (1893–1917)*, tr. J.B. Brough (Dordrecht: Kluwer, 1991).

31. 'The drawn figure and writing are two *sine qua non* conditions of

Stiegler takes this thesis much further, integrating it with Leroi-Gourhan's concept of exteriorisation (see Introduction). Technical objects, for Stiegler, constitute an epiphylogenetic memory, a 'past that I never lived but that is nevertheless my past, without which I would never have had a past of my own'.[32] Epiphylogenetic memory is distinct from both genetic and ontogenetic memory (the memory of the central nervous system); in Stiegler's words, it is a 'techno-logical memory'[33] which resides in languages, the use of tools, the consumption of goods, and ritual practices. We might say then that technics, as the idealisation of geometrical thinking, inscribes time and simultaneously brings into play a new dimension of time—one which, as Stiegler shows, remained under-elaborated in Heidegger's *Being and Time*.

§20.3. GEOMETRY AND COSMOLOGICAL SPECIFICITY

If Stiegler was able to retrieve from his reading of Plato and his deconstruction of Heidegger a concept of time as technics in Western philosophy, it seems that a similar enterprise would not be possible for ancient Chinese philosophy. We have to admit that to say that technology inscribes time is to make an *ontological* and a *universal* claim. Leroi-Gourhan's

---

geometry, qua two dimensions of exteriority. There is geometry only when there is a figure whose elements (point, line, surface, angle, hypotenuse, etc.) are defined by a language that poses them as idealities. But this language can pose them as definitions in this way only on condition that they can be orthographically recorded, allowing the work of thinking to proceed step by step and "to the letter", with no loss of semantic substance.' Stiegler and During, *Philosopher par Accident*, 54.

32.   B. Stiegler, *Technics and Time 1: The Fault of Epimetheus* (Stanford, CA: Stanford University Press, 1998), 140

33.   Ibid., 177

anthropology of technology had already shown that technics should be understood as a form of the exteriorisation of memory as well as the liberation of organs, and hence that the invention and use of technical apparatuses is also a process of hominisation. Tool-use and the liberation of the hands, and the invention of writing and the liberation of the brain, are corresponding activities that transform and define the human as a species. In other words, Leroi-Gourhan offers an evolutionary theory of the human from the perspective of the invention and usage of technical objects. However, the experience of technics is related to and partially conditioned by cosmology—and it is precisely in this sense that we insist on the importance of a *cosmotechnics*. Technical apparatuses function somatically as extensions of organs—and, as prostheses, are *somatically and functionally* universal, and yet they are not necessarily *cosmologically* universal. That is to say, in so far as technics is both driven by and constrained by cosmological thinking, it acquires different meanings, beyond its somatic functionalities alone. For example, different cultures may have similar calendars (e.g. with 365 days in a year), yet this doesn't mean that they have the same concept or the same experience of time.

Now, as touched upon in the Introduction, Leroi-Gourhan himself provides a comprehensive theory of the convergence and divergence of technical inventions across different milieus according to two general concepts: *technical tendencies* and *technical facts*.[34] The *technical tendency* is a universal tendency that occurs in the techno-evolutionary process, for example the use of flint or the invention of the wheel; whereas

---

34.   A. Leroi-Gourhan, *Milieu et Technique* [1943] (Paris: Albin Michel, 1973), 424–34.

technical facts relate to particular expressions of this tendency conditioned by a specific social-geographical milieu: for example, the invention of tools that suit a particular geographical environment or adopt the use of certain symbols.

Yet even if we agree with Leroi-Gourhan in seeing the exteriorisation of memory as a general technical tendency, this does not yet allow us to explain why and how each culture exteriorises at a different pace and with a different direction; that is, it does not explain how exteriorisation is determined by certain conditions—not only biological and geographical, but also social, cultural, and metaphysical. As noted in the Introduction, Leroi-Gourhan attempted to analyse the differences between technical facts in terms of the specificity of the milieu and its exchanges with other tribes and cultures; however, his focus tended to be on the description of technical objects themselves. Indeed, this constitutes the great strength of Leroi-Gourhan's unique research method; and yet in proceeding in this way he failed to take the question of cosmology sufficiently into account.[35] For Leroi-Gourhan, in the differentiation of technical facts the *biological* condition is primary, since it is central to the issue of survival: for example, utensils such as bowls are invented so that one need not go to the source of the water every time. The importance of *geographical* conditions is evident, since the climatic conditions specific to a given region favour certain inventions over others. In *Fûdo*—a response to Heidegger's *Being and Time*—Japanese philosopher Watsuji Tetsurō (和辻哲郎) even goes so far as to argue that the milieu also determines the personal character of the population and

---

35. *Speech and Gesture* does in fact contain passages on the relation between city development and cosmogony—however here Leroi-Gourhan understands the latter as a symbolic form.

their aesthetic judgement.[36] The Japanese word *fûdo* comes from the two Chinese characters for wind (風) and soil (土). Watsuji classifies three types of *fûdo*, namely monsoon, desert, and meadow. To give brief examples of Watsuji's observations, he thinks that, since Asia is heavily affected by monsoons, the resulting relative lack of seasonal change creates an easy-going personality. In Southeast Asia especially, since the weather is always very warm, nature provides a plenitude of foodstuffs, and therefore there is no need to labour too much in order to survive, or to worry about the possibility of day-to-day living. Similarly, he argues that the lack of natural resources in the deserts of the Middle East creates solidarity between peoples, so that the Jewish people, although they live in diaspora, remain united; while in the meadowlands of Europe, clear and regular seasonal changes demonstrate the constancy of the laws of nature, thus suggesting the possibility of mastering nature with science. Watsuji has an interesting observation on the relation between the *fûdo* of Greece and the development of geometry and its logic, as expressed in Greek art and technics. He points out that, far before the sculptor and painter Phidias (480–430 BC), Greek sculpture already had a close relation to Pythagorean geometry. Before the birth of geometry, Greek art already attests to a 'geometrical' mode of observation or *thēoria*, conditioned by a *fûdo* which is 'bright' and 'hides nothing':

> Hence the Greek climate offered a unique opportunity for the furthering of such unrestricted observation. The Greek looked at his vivid and bright world, where the form of everything was

---

36. T. Watsuji, *Climate and Culture: A Philosophical Study* (*Fûdo* [風土]), tr. G. Bownas (Westport, CT: Greenwood Press, 1961).

brilliant and distinct, and his observing developed without restriction in that there was mutual competition [...] The observation of a bright and sunny nature automatically promoted the development of a similarly bright and sunny character in the subject. This came out as a brightness and clarity of form in sculpture, in architecture and in idealistic thought.[37]

Watsuji related this 'pure observation' to Aristotle's concept of form (*eidos*) as essence (*ousia*); one can also relate it to his hylomorphism, and to Plato's theory of form—the incarnation of the ideal in the real. This geometrical reason is crucial to the development of the art and technics that characterised ancient Greek culture. The Romans, although they were unable to take up the legacy of Greek art, maintained their geometrical reason, and therefore, suggests Watsuji, 'through Rome, Greece's reason settled the fate of Europe'.[38] By contrast, in Chinese and Japanese *fûdo* one rarely encounters the brightness of Greece; instead they are characterised by mists and constant changes in weather, meaning that beings are obscured and are not revealed in the same way as in Hellenic forms. According to Watsuji, then, what developed in these *fûdo* was an illogical and unpredictable 'unity of temper':

So the artist, unlike his Greek counterpart, cannot seek unity in his work by proportion and by regularity of form. The latter were replaced by a unity of temper, which cannot be other than illogical and unpredictable. In that it is hard to find any law in them, techniques governed by temper never developed into learning.[39]

---

37.   Ibid., 86.

38.   Ibid., 91. Watsuji was not probably aware that Heidegger had an opposite position on the Greco-Roman heritage.

39.   Ibid., 90.

It is worth mentioning here that Watsuji had already observed that the *fûdo* is not eternal: he predicted that the situation in South East Asia would change greatly when Chinese businessmen entered into the region—meaning that the technologies, practices, and social values brought by the Chinese through trade would produce a huge transformation of the region. It is only in so far as exchange between ethnic groups is limited that the conception and development of technics is subsumed by cosmology, which is grounded in culture, social structure, and moral values—and, for Watsuji, ultimately, in the *fûdo*.

The fact that Chinese culture does not elaborate on time and geometry, then, may have served as a cultural and cosmological condition of its technological development, producing, in Leroi-Gourhan's terms, different technical facts within the universal technical tendency. We can observe the different ways in which these conditions developed in China and in the West according to two technical aspects: firstly, in the interpretation of time in the production of technical beings, in the sense that time can be geometrically treated, whether as linear or cyclical, thus allowing a new temporalization; and secondly, in the understanding of progress and historicity in relation to technicity. These differences stem from differing understandings of nature (cosmos) and progress (time). In *Procès ou Création*, his treatise on the Neo-Confician Wang Fuzhi (王夫之, 1619–1692), Jullien remarks that Wang Fuzhi can hardly speak of the progress of history when he does not oppose nature to history; Jullien's conclusion is that 'the tradition in which [Wang] inscribed his thinking was never affected by a theophanic reading of history'.[40] But there is also

---

40.   Jullien, *Procès ou Création*, 72.

a *political* reason that explains why, although there is history in China, there is also a lack of discourse on historicity. It is surprising to note that Laozi, the author of *Dao De Ching*, was a historian of the Zhou dynasty; or more precisely, that he was the official historian of the royal library.[41] What did it mean to be a historian at that time? And how can a historian have left us a *Dao De Ching* that is indifferent to both history and time? In its first sentence we already read that '*Dao* called *dao* is not *dao*. Names can name no lasting name'. Should we read this as a refusal of the writing of history, provided that what he means by history is something that always escapes and is continually changing? In fact, in Laozi's time, the role of the historian was to peruse the ancient texts in order to give advice on governance; the political use of history qua interpretation of texts took precedence over any development of historical consciousness. As we have already seen in Part 1, this remained the case up until the time of Dai Zhen, and especially Zhang Xuecheng, who, in the eighteenth century, sought to break *Dao* out of the 'prison' of the classics.

This second point will be our major concern, meaning that we will be concerned with articulating the relation between the conception of time (nature and history) and technological development. In parallel, we will see that the efforts that were made to elaborate on the concept of time in China and in East Asia were in general closely related to the question of modernity, but that they adopted a very ambiguous relation to technics. This has its consequences in contemporary China, where today we observe a kind of paradox: on one hand, there is rampant

41.   Ssu-Ma Ch'ien (司馬遷), *Records of the Grand Historian of China* (史記), tr. B. Watson (New York: Columbia University Press, 1961); see the section on Lao Tzu.

technological development in terms of scientific research, infrastructure projects, and construction (including its development project in Africa); while on the other hand there is a strong sense of loss or disorientation, in which China ceases to be China, and becomes instead capitalism with Chinese characteristics—not so different from the situation foreseen by Hu Shi (see §16.2), where the residue of Chinese culture serves only to inflect an otherwise triumphant westernisation. The end of modernity in Europe, meaning the beginning of the process of gaining technological consciousness, has only amplified this paradox, since the temporal and spatial compression of globalisation leaves no room for negotiation, only exerting an ever greater pressure for assimilation.

This is a delicate hypothesis, and so is the demonstration that I want to proceed with in the following pages. The aim here is to reconsider the question of technology by situating China within the European temporal axis; and to make room for a new programme of cosmotechnics. However, we will first have to examine the different attempts to 'overcome modernity' and to learn from their failures. These historical lessons are indispensable in exposing the deep problematics of modernity and the traps that may lie ahead as we try to move beyond them.

§21. MODERNITY AND TECHNOLOGICAL CONSCIOUSNESS

If, as we saw in Part I, the holistic cosmological view in China was brutally dismantled by modernisation, this was because it was able neither to resist nor to confront the technical reality of European and American culture. *Qi-Dao* as a moral and cosmological structure was transformed and restructured by the material-ideal structure of technics. The sun, moon, and planets moved in the same way as before, but they were no longer perceived to have the same meaning, the same structure, or the same rhythm. Modernisation is fundamentally a transformation, if not a destruction, of the moral cosmology that is expressed in every form of art in China, from tea ceremonies to calligraphy, from craftsmanship to architecture.

After the example of Plato's suppression of the spatial supplement involved in the slave-boy's anamnesis, technics as inscription, and hence as a support of time, has been the unconsciousness of the modern. That is to say, it has never been thematised as such within modernity, and yet it acts in such a way as to constitute the very conception and perception of the modern. Now, unconsciousness only exists in relation to consciousness; we might even call it the negation of consciousness. When consciousness recognises something unconscious, even though it may not be able to know exactly what it is, it will attempt to integrate it and render it functional. Technological unconsciousness is the most invisible, yet the most visible being; as Heidegger says, we don't see what is nearest to us. And it was this technological unconsciousness that granted the *cogito* the will and the self-assurance to exploit the world, without perceiving the limits of this exploitation. The later discourse on progress and development that fuelled and justified the European colonial project continues with the same logic, up until the moment when crises are imminent: industrial catastrophes, the extinction of species, the endangering of biodiversity....

Bruno Latour formulates this in a different way: he sees it as the internal contradiction between two registers: on the one hand what he calls 'purification', e.g. nature vs culture, subject vs object, and on the other hand what he calls 'mediation' or 'translation', meaning the production of 'quasi-objects', or objects that are neither purely natural nor cultural (for example, the hole in the ozone layer). The latter, presented as a hybridisation, are according to Latour in fact nothing but the amplification of purification. Given this contradiction in the constitution of the modern, Latour claims that 'we have never been modern', in the sense that the 'modern' profoundly separates nature and culture, and embodies the contradiction between domination and emancipation. Although Latour does not characterise the modern in terms of technological unconsciousness, then, he recognises that the modern refused to conceptualise quasi-objects. A quasi-object is something that is neither merely object nor subject, but a technical mediation between the two—for example (in Michel Serres's example) a football in the football game which, when the two teams play, ceases to be an object, but transcends such a subject-object division. The refusal to conceptualise quasi-objects means that the concept of technics that functions to separate nature and culture, subject and object, as is the case in the laboratory, is not fully recognised or remains unconscious:

> Moderns do differ from premoderns by this single trait: they refuse to conceptualize quasi-objects as such. In their eyes, hybrids present the horror that must be avoided at all costs by a ceaseless, even maniacal purification […] [T]his very refusal leads to the uncontrollable proliferation of a certain type of being: the object, constructor of the social, expelled from the social world, attributed to a transcendent world that is, however, not divine—a world that produces, in contrast, a floating subject, bearer of law and morality.[42]

42.   B. Latour, *We Have Never Been Modern*, tr. C. Porter (Cambridge, MA: Harvard University Press, 1993), 112.

Technics remained unconscious, then, and yet this unconsciousness began to produce significant effects in the life of the mind at a certain moment in European history, namely the modern era, and this unconsciousness culminated during the Industrial Revolution. The transformation of this unconsciousness to consciousness characterises the contemporary technological condition. It is a *turn*, in which one attempts to render technics a *part* of consciousness, but not consciousness itself (which is why we can understand it as instrumental rationality). This new condition is shared throughout the globe, without any choice in the matter: even in the Amazonian forest there are movements that have had to insist on their own cultures—for example giving rights to non-humans, preserving traditional cultural practices, and so on—just as the Chinese attempted to save their traditional values during modernisation. Faced with the impossibility of claiming complete social and economic autonomy, they have had to confront the contemporary technological condition, and the destiny of such indigeneous practices remains uncertain today.

In contrast to the common understanding, according to which the postmodern, dating to the late twentieth century, indicates the end of modernity, I would rather say that modernity only comes to an end at this moment in the twenty-first century, almost forty years after Jean-Francois Lyotard's announcement of the advent of the postmodern, since it seems that only at this stage do we come to appreciate our technological consciousness. In fact, not only Latour and Lyotard, but also many others who wrote on technology, such as Jacques Ellul and Gilbert Simondon, had raised the problem of the lack of awareness and misunderstanding of technology. For example, in *On the Mode of Existence of Technical Objects*, Simondon characterises it as an ignorance and misunderstanding of technics, and tries to render visible or raise awareness of technical objects.[43] Jacques Ellul, in turn, took up Simondon's analysis of technical objects and technical ensembles and extended it to the global technological system that is in the process of becoming a totalising force. It is this effort at rendering conscious that of which we are unaware, but which largely constitutes our everyday life, that really constitutes the 'end of modernity'.

However, let us step back and ask: *What do we mean by the word 'end'?* It doesn't mean that modernity suddenly stops, but rather that, as a project, it has to confront its limit, and in doing so, will be transformed. Therefore, by 'end of modernity', we surely do not mean that modernity ceases to affect us, but rather that we see and know that it *is* coming to its end. Nevertheless it still remains for us to overcome it, to overcome the effects that it has produced on and in us—and this will undoubtedly take much longer than we might imagine, just as Heidegger tells us that the end of metaphysics doesn't mean that metaphysics no longer exists and has ceased to affect us, but rather that we are witnessing its completion and waiting for something else to take over, whether a new thinking of Being, or an even more speculative metaphysics. Furthermore, like the end of metaphysics, the end of modernity proceeds at a different pace in Asia than in Europe, precisely in so far as, firstly, their philosophical systems do not perfectly match each other, and, secondly, the propagation of a concept from one system to another is always a deferment and a transformation.

The postmodern arrived too early, with too much hope, anxiety, and excitement, in the writings of Lyotard—a prophet of the twentieth century. Lyotard's discourse on the

---

43.   Simondon, *Du mode d'existence des objets techniques*, 10.

postmodern prioritises the aesthetic; he is sensitive to the aesthetic shifts produced by the transformation of the world driven by the forces of technology, and tries to turn this force into one that would be capable of negating the modern. The postmodern is a response to such new aesthetics, and also serves as a new way of thinking through the appropriation of technology. Hence it is unsurprising to see that, in the exhibition *Les Immatériaux* curated by Lyotard in 1985 at the Centre Pompidou in Paris, the question of *sensibility* came to the fore, with new technical and industrial objects juxtaposed with artworks from Yves Klein, Marcel Duchamp, and others. The sensibility and the 'inquietude' that *Les Immatériaux* attempted to foster amounts to an awareness of the uncertainty of the cosmos, the insecurity of knowledge, and the future of humanity. With this new sensibility, human beings become more aware of what is in their hands, of the technical means they have developed, and of the fact that their own will and existence have become dependent upon these apparatuses which they believed to be their own creations—and indeed that the human itself is in the process of being 'rewritten' by the new 'immaterial' languages of machines. It is in this way that Lyotard raised the question of anamnesis in relation to technology: he saw very clearly that the exploitation of memory by industry would be amplified with the development of telecommunications technologies. He therefore sought to overcome the industrial hegemony of memory by pushing the question to a new height (and setting it on a new plane), albeit one that remains very speculative and hence almost opaque. The process that is understood as the end of modernity, in my own conceptualisation, centres on the hypothesis that modernity is subtended by a technological unconsciousness, and that its end is indicated by a becoming-conscious, a realisation that Dasein is a technical being who may invent technics, but is also conditioned by them. Heidegger's *Being and Time*, especially his critique of Cartesian ontology, and in his later works the effort to reconstruct the history of Being—a task which can be understood as that of terminating modernity by posing a new question, a recommencement—arises from an awareness of the forgetting of Being. The ontological difference is an opening, since it reformulates the question of Being according to two different orders of magnitude, one concerning beings (*Seiendes*), the other Being (*Sein*). The forgotten question of Being functions as the unconscious of the ontic inquiry into beings constituted by the history of science and technology. Freud, in turn, developed a theory of the unconscious and of repression in order to retrieve that which is deeply hidden and long since forgotten and repressed by the superego. The tasks of Freud and Heidegger, although they belong to two very different theories and disciplines, characterised two major discourses on modernity in the twentieth century, and two attempts to quit this modernity. As we shall see, in confronting the question concerning technology in China, Freud's conception of the unconscious, repression, and working-through will be crucial. Indeed, Heidegger hinted at a kind of repression inherent in the antagonistic relation between technology and the question of Being: for him, technology, the *completion* of Western metaphysics, occluded the original question of Being. The forgetting of Being, in effect, *is* the question concerning technology. In order to understand technology, and what is at stake in it for non-European cultures, then, we must go by way of Heidegger and the concept of technology as the completion of metaphysics, but without equating Eastern and Western philosophical systems and thereby attributing a universal origin of technics to Prometheus. We must rather seize the possibility of *appropriating* it, *deferring* it as an end, and, in this deferring, re-appropriate the *Gestell*—that is, modern technology.[44]

It is with Bernard Stiegler, not Lyotard, that this question becomes *transparent*. The work of Stiegler announces the end of modernity.[45] Stiegler demonstrates that Western philosophy has long since *forgotten* the question of technology: if, for Heidegger, there is a forgetting of Being, for Stiegler there is equally a forgetting of technics. Technics, as tertiary retention, is the condition of all conditions, meaning that even Dasein, who seeks to retrieve an authentic time, in order to do so must rely on tertiary retention, which is at once the already-there and the condition of Dasein's being-in-the-world. For Stiegler, technics, notwithstanding its destructive nature in the epoch of technology described by Heidegger in 'The Question Concerning Technology', thus becomes more fundamental than the forgetting of Being: the history of Being as situated in the history of Western metaphysics will have to be rewritten according to the concept of technique as an original default (as well as the fault of Epimetheus).

We may therefore ask whether, as suggested above, rather than this forgetting being a lack of memory, a *hypomnesis*

---

44. This is derived from what Stiegler, following Derrida, calls 'pharmacology', meaning that technics is at the same time 'remedy' and 'poison'. We will see later that the resistance that we are talking about is by no means a blind resistance to all modern technologies—which would be unwise if not impossible—but rather a resistance that aims for a re-temporalization and re-opening of the question of world history.

45. This is not to suggest that Lyotard did not contribute to it. As we will see below, Lyotard posed a very speculative question in a discussion with Stiegler—that of the 'clear mirror'—which attempts to radically open up a new direction in the dialogue with the Other that is often absent in the philosophy of technology.

brought about by technical objects, it is a question of an *unconscious* content that is only slowly recognised once its effects on the life of the mind become significant. The deconstruction of Heidegger's and Husserl's concepts of time in the three volumes of *Technics and Time* may in this case be seen as a psychoanalysis of this technological unconsciousness, and therefore as an attempt to release technics from their repression by the *cogito,* the symbol of modernity.

§22. THE MEMORY OF MODERNITY

Stiegler's tertiary retention is fundamentally a question of a kind of time that remains ambiguous in Heidegger's *Being and Time*. Heidegger's critique of clock time forms a part of his critique of the forgetting of Being, as indicated by the loss of an authentic time, or *Eigentlichkeit*. In the second division of *Being and Time*, Heidegger expanded this critique to encompass the question of history and historicity. In order to understand historicity, one must first situate Dasein as a historical being. Heidegger distinguishes historicity (*Geschichtlichkeit*), which has its source in Dasein's historising (*Geschehen*), from historiology (*Historie*): historicity is not an objective description of what is past, but rather resides in the totality of historising, meaning the temporalisation of the past, present, and future. For Heidegger the past, memory, is primordial, as is the case in Wilhelm Dilthey—a major influence on Heidegger both before and during the writing of *Being and Time*. For Dilthey, life is historical in three basic ways. Firstly, the past always insists in the present, since life is always an *Innewerden*, a process of integrating what is past into the present; secondly, the present is a building-up (*Aufbau*) of the past in terms of structure and development; and thirdly, the past also exists as an objectified past, in the form of artefacts, nexuses of actions, events, and so on.[46] Not unlike Dilthey, Heidegger attempts to grasp this temporalisation as a whole. The present, as the pivot of such a historisation, emerges from Dasein's grasp of its own historicity.

In the second division of *Being and Time*, Heidegger arrives at the question of resoluteness, being-towards-death, and being-in-the-world as the basic structure with which to describe this temporalisation which yields an 'authentic historicity'. The world is disclosed in the resoluteness of Dasein, since in resoluteness Dasein comes back to itself; and in such a coming-to-itself, it is able to find its authenticity. But what does Heidegger mean by resoluteness (*Entschlossenheit*)? It is, Heidegger writes, defined

> as a projecting of oneself upon one's own Being-guilty [...] Resoluteness gains its authenticity as *anticipatory* resoluteness [*als das verschwiegene, angstbereite Sichentwerfen auf das eigene Schuldigsein* [...] *Ihre Eigentlichkeit gewinnt sie als* vorlaufende *Entschlossenheit*]'.[47]

Derrida points out that here, the *Schuld* of the *Schuldigsein* doesn't simply mean guilt (*coupable*) or responsibility, but rather a non-empirical debt, 'of which I am indebted as if I was always already taken in a contract—and it is historicity—a contract that I have not signed but that ontologically obliges me.'[48] This 'non-empirical debt' is the 'heritage' whose

---

46.   T.R. Schatzki, 'Living Out of the Past: Dilthey and Heidegger on Life and History', *Inquiry* 46:3 (2003), 301–323: 312.

47.   M. Heidegger, *Sein und Zeit* (Tübingen: Max Niemeyer Verlag, 2006), 382; *Being and Time*, tr. J. Macquarrie and E. Robinson (Oxford: Blackwell, 2006), 434.

48.   J. Derrida, *Heidegger: la question de l'Être et l'Histoire. Cours de l'ENS-Ulm (1964–1965)* (Paris: Galilée, 2013), 273–4.

authenticity can only be achieved when Dasein firstly takes over 'in the thrownness an entity which is itself'.[49] Such resoluteness, in turn, is brought about by the recognition of being-towards-death as finitude and limit of Dasein. In other words, being-towards-death is the necessary condition of any freedom in its 'authentic sense'. Only in being free for death does Dasein understand its finite freedom, which allows it to choose and to decide between accidental situations, and therefore become able to hand down to itself its own fate. This self-handing-down (*sich überliefern*) of resoluteness must result in the revelation of a place, the '*da*' or '*there*' of Dasein, as its destination in authenticity. In what does this self-handing-down consist, then?

> The resoluteness in which Dasein comes back to itself, discloses current factical possibilities of authentic existing, and discloses them *in terms of the heritage* which that resoluteness, as thrown, *takes over*. In one's coming back resolutely to one's thrownness, there is hidden a *handing down* to oneself [*sichüberliefern*] of the possibilities that have come down to one, but not necessarily as having thus come down.[50]

The *Sichüberlieferung* doesn't happen naturally, but is both a choice and a repetition. Derrida translates it as 'auto-transmission' and 'auto-tradition', and proposes that it is another face of the 'auto-affection' of pure time that Heidegger described in *Kant and the Problem of Metaphysics*.[51] The 'there' is

---

49.   Heidegger, *Being and Time*, 434.

50.   Ibid., 434 [emphasis in the original].

51.   Derrida, *Heidegger, la question de l'Être et l'Histoire*, 265–8. Unfortunately Derrida didn't fully develop his argument, but he does point out that the *sichüberlieferung*, the transmission of the self ('*la transmission de soi*') is the original synthesis and is central to historicity.

revealed in the moment of vision (*Augenblick*), where Dasein resolves the tension between its resoluteness and its being-in-the-world with others. Despite the fact that the question of 'heritage' is recognised, it is recognised only as the 'given'.

But can a historical being be possible without an *analytics* of the 'already there' (*schon da*)? Death only acquires its meaning when it is situated within a world of symbols, relations, and writings; otherwise the death of the human being would be no different to that of animals. Death for animals is fundamentally a question of survival, but for human beings, it is also, according to Heidegger, the question of freedom. It is this question—the question of a *Dasein analytics* from the perspective of technics—that Stiegler attempts to answer in his *Technics and Time*. For Stiegler, temporalisation is conditioned by tertiary retention since, in every projection, there is always a restructuring of memory that is not limited to the past that I have lived. Addressing the museum of antiquities, Heidegger asks 'What is the past?', and replies, 'nothing else than that world within which they belonged to a context of equipment and were encountered as ready-to-hand and used by a concernful Dasein which was-in-the-world'.[52] The past consists in the structures of relations which are no longer expressed as the ready-to-hand, but can only be made visible by thematisation (in which case, they become present-at-hand). However, with the notion of tertiary retention introduced by Stiegler, what *was* ready-to-hand functions as the condition for, and as the unconscious part of, our everyday experience. That is to say, Stiegler brings about a new dynamic of temporalisation; we will come back to this point below, in the discussion of Keiji Nishitani's interpretation of Heidegger.

---

52. Heidegger, *Being and Time*, 431.

The question of memory does indeed concern tertiary retentions such as monuments, museums, and archives: the latter become the symptoms of technological unconsciousness because, on the one hand, this technological unconsciousness speeds up the destruction and disappearance of traditional life, yet on the other also promotes a desire to retain what is disappearing. This is a contradiction, since this memorialisation tends to act as a consolation for the profound melancholia produced by this process, without realising that it is technological unconsciousness that is responsible for it. Modernity, fully dominated by its will, sees only its destination (development, commerce, etc.), and rarely sees what is unconsciously driving it towards such an illusory goal. Hence modernity and memory sometimes seem in opposition to each other, yet at other times seem to supplement each other. The force of modernity is one that dismantles obstacles and abandons laggards, and the critique of modernisation has often centred on its disrespect for history and tradition. Yet the discourse of collective memory is also wholly modern—a compensation for what is destroyed, since only when threatened by destruction does it become a memory rather than a mere object of everyday life of interest only to historians.[53] Heidegger takes a critical view of this memorialisation, since it objectifies in a way that tends to alienate Dasein's authentic grasp of historicity. Objectified history, or what Heidegger calls 'historicism' (*Historismus*), has its source not in Dasein but rather in the effort to objectify world history, in which Dasein then becomes no longer a historical being, but rather one

---

53.   For this reason we see that, in China, high-speed economic development has destroyed cities yet, at equal speed, replaced them with monuments or museums; one suspects that this is not purely an economically driven process, but rather that there is a symptomatic lack of historical consciousness here, as we discuss below.

among many objects, swept along by a history determined by events external to it. In the *Black Notebooks*, Heidegger makes the opposition even more explicit:

> History [*Historie*]: technology [*Technik*] of '*Geschichte*'
> Technology [*Technik*]: history [*Historie*] of 'Nature'[54]

We can understand the statement 'technology is the history of nature' as affirming that technology, identified with the history of metaphysics, underlies the process of the objectification of nature; in the same way, '*Historie*' becomes metaphysical, and conceals *Geschichte*. This opposition is taken up again in the *Black Notebooks* when Heidegger writes:

> Presumably the historian understands '*Geschichte*' as *Historie*, then it is indeed, as he says hypothetically. *Historie* is only a form of technics in the essential sense [...] Only when the power of *Historie* is broken does *Geschichte* again have its space. Then there is fate and openness for the appropriate [*Schickliche*].[55]

But this tension set up by Heidegger between history (*Historie*) and historicity (*Geschichtlichkeit*) can only be resolved when, as in Stiegler, one affirms that the latter cannot do without the former; which also implies that the authenticity of Dasein is always, in a sense, inauthentic—that is, deprived of any absoluteness or certainty. Modernity only ends, and historicity (albeit in a different sense to Heidegger's) is only achieved, when the question of memory is rendered

---

54. M. Heidegger, *GA 95 Überlegungen VII-XI Schwarze Hefte 1938/39* (Frankfurt am Main: Klostermann, 2014), 351.

55. Heidegger, *GA 97*, 29.

transparent, meaning that technological unconsciousness is rendered into a memory—a memory whose significance and impact one must become aware of.

The end of modernity is therefore indicated not only by the acknowledgement that the human being is no longer the master of the world, or that the world escapes us. This we have known since the very beginning of humanity: the Gods were above us, no matter whether they were those of Mount Olympus, of Egypt, or of the Sinai Peninsula; since the very beginning it has been known that the notion of the human as master of the world is only an illusion; but at the moment when this illusion is fuelled by technological unconscious, it begins to structure reality itself. The end of modernity is a re-cognition of this illusion; the recognition that technics is what conditions hominisation, *not only in its history but also in its historicity*. The end of modernity therefore consists not only in the enunciation of this end, but also in a reformulation of the history of Western metaphysics, as in Nietzsche's *Gay Science*, where the madman crying incessantly in the marketplace searches for the lost god.[56] The transcendence of God will have to be replaced either by a philosophy of immanence or by another transcendence—the transcendence of Being[57] and Dasein.[58]

---

56.   Nietzsche, *The Gay Science*, 119–20 [§125].

57.   '*Sein ist das transcendens schlechthin* (*Being is the* transcendens *pure and simple*)' Heidegger, *Sein und Zeit*, 38 [§7] [italics in the original]; Dermot Moran has remarked that Heidegger took this point up again in 'Letter on Humanism': 'This retrospective definition of the essence of the Being of beings from the clearing of beings as such remains indispensable for the prospective approach of thinking toward the question concerning the truth of Being'. See D. Moran, 'What Does Heidegger Mean by the Transcendence of Dasein?', *International Journal of Philosophical Studies*, 22:4 (2014), 491–514: 496.

58.   In the same passage (§7) of *Sein und Zeit*, 38 (62), Heidegger writes: '[...] the transcendence of Dasein's Being is distinctive in that it implies the possibility and the necessity of the most radical *individuation*. Every disclosure

Stiegler adopts Derrida's method so as to reconstruct a history of technics as an onto-epistemological object. This is undoubtedly an ambitious project: Stiegler wants to re-read the history of philosophy through technics, and hence to make technics the first question of philosophy. In his reformulation of the mythology of Prometheus, fire constitutes the origin of man as a technical being. Recall that Zeus commanded Prometheus to distribute skills to all living beings, including humans and animals. Epimetheus, the brother of the giant, proposed to take over this task. But Epimetheus—whose name means 'hindsight' in Greek—forgot to distribute any skill to humans, and hence Prometheus had to steal fire from the god Hephaestus. The punishment Prometheus received from Zeus was to be chained to the cliff, while the *Aetos Kaukasios* ('Caucasian eagle') came to eat his liver every day after it had regrown during the night. A human without fire—meaning without technics—would be an animal without quality. Its origin is the default; hence Stiegler proposes to think of this default as a necessity (a '*défaut qu'il faut*'). In Stiegler's reinterpretation, the myth of Prometheus and Epimetheus lies at the centre of classical Greek thought, and constitutes the unconscious of Western philosophy.

Hence, for Stiegler, the history of Western philosophy can be also read in terms of the history of technics, in which the

---

of Being as the *transcendens* is *transcendental* knowledge. *Phenomenological truth* (the disclosedness of Being) is *veritas transcendentalis*'. For a more comprehensive discussion on Heidegger's concept of the transcendence of Dasein in relation to the Husserl's phenomenology, see Moran, 'What Does Heidegger Mean by the Transcendence of Dasein?'. It suffices here to mention that in *What is Metaphysics?*, Heidegger writes: 'Da-sein means: being held out into the nothing [*Hineingehaltenheit in das Nichts*]. Holding itself out into the nothing, Dasein is in each case already beyond beings as a whole. This being beyond beings we call transcendence [*Dieses Hinaussein über das Seiende nennen wir Transzendenz*]'. Cited by Moran, 508.

question of Being is also the question of technics, since it is only through technics that the question of Being is opened to us. A similar reading of Heidegger was proposed by Rudolf Boehm in his 1960 essay 'Pensée et technique. Notes préliminaires pour une question touchant la problématique heideggerienne [*Thinking and Technics: Preliminary Notes for a Question Regarding the Heideggerian Problematic*]' which, as we touched upon in Part 1, concerns the interpretation of Heidegger's 1935 *Introduction to Metaphysics*. Boehm shows that *technē* is always present not only in Heidegger's thinking, but also as the ground of Occidental philosophical thinking. Indeed, it is this technics that characterises the metaphysical mission of the Ionian philosophers. Boehm shows that, in *Introduction to Metaphysics*, Heidegger interprets technics in the Ionian philosophers as an activity that produces a radical opening of Being through the confrontation between *technē* (that of the human) and *dikē* (that of Being). We have attempted to recover the concept of *technē* in Presocratic philosophy in Heidegger's *Introduction and Metaphysics*, and we have seen (§8) how Heidegger translated *dikē* not as *Gerecht* (justice), but rather as *Fug* (fittingness); In war (*pōlemos*) or strife (*eris*), Being reveals itself as *physis*, *logos*, and *dikē*.[59]

For Heidegger, however, this reading of technics as the origin of philosophical and practical activities that opens the question of Being is foreclosed in Platonic-Aristotelian Athenian philosophy as a declension (*Abfall*) and a fall (*Absturz*)[60]—the beginning of onto-theology. According to Boehm's reading, Heidegger believes that Plato and Aristotle opposed technics to nature, and therefore excluded technics from its original

---

59.   Backman, *Complicated Presence*, 33.

60.   Boehm, 'Pensée et technique', 202.

meaning, as developed by the Ionian philosophers (an omission that Stiegler undertakes to correct). For Heidegger, then, if the danger of modernity consists in the rise of technology, this technology is essentially different from the *technē* of ancient times. Technological development, accompanied by its rationality and driven by the desire for mastery, forms a gigantic force that is in the process of depriving the world of any other possibility and turning it into a giant standing reserve, as *adikia* or *Unfug* (un-fittingness).[61] Technology is the destiny of Western metaphysics, and indeed this is even clearer when we recall Heidegger's famous assertion that 'cybernetics is the completion or the "end" of metaphysics'.[62] The question here is not that of judging whether or not this critique is just, but rather that of seeing it as a contribution to a movement away from the technological unconsciousness of the modern. Toward the conclusion of his essay, referring to the necessary confrontation between *technē* and *dikē*, Boehm raises two very intriguing questions:

> [C]ould philosophy not forget Being, and simply concentrate all of its efforts on attaining the highest perfection of its technics? Or else, ultimately, is there any possibility that thinking could release itself from its attachment to a technical condition?[63]

We can identify Boehm's two questions with two forms of thinking which today confront modernity: one seeks to overcome the impasse of philosophy analysed by Heidegger

61. See our discussion in Part 1 (§8).

62. M. Heidegger, 'The End of Philosophy and the Task of Thinking', in *On Time and Being*, tr. J. Stambaugh (New York: Harper & Row, 1972), 55–73.

63. Boehm, 'Pensée et technique', 217.

through a new conceptualization of technology, as is the case in Stiegler; the other tends to retreat into a 'philosophy of nature', whether Whiteheadian or Simondonian—to submit *technē* to nature—namely, to surrender to the overwhelming (*Überwaltigend*), or Gaia. We have already touched on the limits of this second approach in the Introduction: Chinese philosophers such as Mou Zongsan, and Sinologists such as Joseph Needham, have already discovered the affinity between Whiteheadian and Chinese philosophy; but if we are to admit that a return to the Whiteheadian concept of nature can help us to escape the impasse of modernity, then would a return to the Chinese traditional philosophy also afford such an escape route? Maybe we should ask the same of indigeneous ontologies: Are they then able to confront technological modernity? Our task here is to show that this is not sufficient. In the case of China, the *Qi-Dao* unity has been completely shattered. Although one may wish to argue that because of the formidable political factors in play, we cannot give an absolute or negative answer to this question, our philosophical analysis in Part 1 concerning the breakdown of the *Qi-Dao* relation, and our analysis above of the geometry-time-technics relation in China in comparison to Europe, have aimed to show that this is not only a socio-political question, but fundamentally an ontological one. Those who propose a return to nature or to cosmologies alone seem to have gracefully elided the failures of the project of 'overcoming modernity' in the twentieth century. These failures must be addressed. As will become evident below, the Kyoto School's fanatical attempt to carry through this project, for example, is something we should avoid at all costs today, but their analyses of the question of time and historical consciousness remain important in posing anew the question of technology and world history.

§23. NIHILISM AND MODERNITY

As stated above, the long process known as 'modernity' in Europe did not take place in China or in other Asian countries. The mastery of the world as a will to power did not emerge in China,[64] and technological unconsciousness, since it produced such a negligible effect, was never considered as a problem to be overcome. As we saw in Part 1, technology only became a problem following the Opium Wars. But is the China of today ready to take up the question of technology and to give it sufficient reflection from the perspective of its own culture and tradition? Because even today, if we are to take up Heidegger's and Stiegler's critiques, we risk accepting a universal history of technology and a cosmopolitanism *without* world history.

This risk is reflected in current thinking on the opposition between global and local. In such an opposition, the local is seen as a form of resistance against the global; yet the discourse of the local is itself the product of globalisation. The fundamental necessity is that the relation between technics and time should be further examined, not so as to dismantle its ontological ground, which the European philosophers have already laid bare, but rather so as to understand its implications for cultures in which such a reflection has not yet occurred; and further, to develop a new programme that does not consist in a retreat into the local, the 'uncontaminated', whether as resistance or as passive adaptation to the global. Indeed, here we should

---

64. We should count the Yuan dynasty, governed by the Mongolians, as an exception, since China was invaded and colonised.

also question the image of the globe, which intuitively suggests that modernisation and de-modernisation is a spatial question subject to the logic of inclusion and exclusion. For this reason, below I will propose instead to think from the perspective of a global axis of *time*.

Keiji Nishitani, of the Kyoto School, was one of the few early twentieth-century Asian philosophers who formulated a profound philosophical critique of technology in relation to the question of time. This is not surprising when we consider that Nishitani was once a student of Heidegger in Freiburg, and like his teacher, he was also associated with fascism in Japan, and was consequently suspended from teaching after the Second World War. His understanding of technology—and here we resist passing judgment, since what is said below should prepare us to grasp the common root of their metaphysical fascism—resonated with Heidegger's critique of modern technology, but whereas Heidegger looked to the early Greeks, Nishitani attempted to propose a 'solution' from and for the East.

In his early work, Nishitani set himself the task of showing how, unlike Western philosophy, Eastern philosophy was able to transcend nihility, or more precisely to demonstrate such a possibility by appropriating the categories of Western philosophy. What exactly is this nihility that seems to Nishitani to be a kind of divide between the two systems of thought?

Nihility refers to that which renders meaningless the meaning of life. When we become a question to ourselves and when the problem of why we exist arises, this means that nihility has

emerged from the ground of our existence and that our existence has turned into a question mark.[65]

Nihility is like an evil that arises in every questioning of existence. There are two instances in which it becomes possible to
ignore it: either in the permanent objectification of the world,
in which the question of subjectivity becomes a non-question
and yet a gigantic force pushes human beings towards the
abyss of nihility; or in a system of thinking that offers a remedy
for the emergence of nihility—not just resisting it, but casting it
into an absolute emptiness which Buddhists call 'the void [空]'.
According to Nishitani, modern science and technology
are accelerating humanity towards a situation in which the
question of Being presents itself as a crisis. Reflecting, like
Heidegger, on the relation between science and technology, Nishitani argues that science consists in universalising
the laws of nature, since they are regarded as the absolute
and most objective rules; hence they can enter into realms
wherein they were previously considered irrelevant or illegiti-
mate as explanatory means. These supposedly universal laws
of nature are implemented in technology, and therefore their
effects are amplified not only in the natural realm but also in
the social and economic realms. Two consequences derive
from this: firstly, the laws of nature pervade every domain;
and secondly, their impact is *amplified* by technology in such
a way that they can assert power outside of their own realm:

> [T]hrough the work of man [...] the laws of nature become mani-
> fest in their most profound and obvious mode. In the machines,
> human work can be said to have passed beyond the character

65.   Nishitani, *Religion and Nothingness*, 4.

of human work itself, to have objectified itself and to have assumed the character of an immediate working of the laws of nature themselves.[66]

The laws of nature, according to Nishitani, are abstractions, in that 'they are nowhere to be found in the world of nature';[67] and yet the world comes to be reconstructed according to these abstractions, thus converting the real to the ideal. Modern technology, which embodies the laws of nature, therefore liberates them from nature itself. According to Nishitani, this dialectical movement has two further consequences: firstly, on the side of man, it produces 'an abstract intellect seeking scientific rationality'; and secondly, it produces a 'denaturalised nature' which is 'purer than nature itself'.[68] The technologised world is thus constructed according to an untruth which follows neither human nature nor nature itself. This opens up a ground for nihility, since man believes only in the laws of nature, which distance him from nature and truth; and the rules of nature thus implemented in technology and embedded in everyday life produce a second distancing of man from truth. French Existentialism, largely influenced by Heidegger, seems to Nishitani insufficient to deal with the situation, since its desire is 'inherent in a nihilism that has yet to seek self-consciousness', and therefore it cannot tackle the root of nihilism[69]—meaning that Sartrian existentialism is still rooted in the Western tradition, especially that inspired by Heidegger, so that its reflection on nihilism doesn't go to the root of

---

66.   Ibid., 83.
67.   Ibid.
68.   Ibid., 85.
69.   Ibid., 88

the problem. The history of Being discussed by Heidegger and Nietzsche, according to Nishitani, 'doesn't exist in the East'. Yet he goes on to argue that 'the East has achieved a conversion from the standpoint of nihility to the standpoint of *sūnyatā*', thus transcending what Hegel calls a 'bad infinity' (*schlechte Unendlickeit*):[70]

> In Buddhism, true transcendence, detached from the 'world' of *Samsāra* as such, has been called nirvāna [...] *Nirvāna* converts the *schlechte Unendlichkeit* into 'true infinity' [...] that is, away from finitude as 'bad infinity' in *Existenz* to infinity in *Existenz*.[71]

There are two questions that interest us in this statement: (1) How is this 'conversion' from the bad infinity to the true infinity possible? and (2) What does it mean in relation to historicity, and world historicity? Nishitani's grasp of the *sūnyatā* is based on a new logic, one that revokes the 'excluded middle'— meaning that it is neither affirmative nor negative. We might call it a privative logic (non-being), in between affirmation (being) and negation (not-being). For Nishitani, science and technology are based on a substantialist thinking that seeks to grasp the essence of being as self-identity. In this reading, developed on the basis of Dōgen's teaching, the logic runs as follows: In order to *be* without self-identity, one has to negate both one's own negation and affirmation. As Dōgen says: 'just understand that birth-and-death is *nirvāna* [...] only then can

---

70. G.W.F. Hegel, *The Encyclopaedia Logic*, tr. T.F. Geraets, W.A. Suchting, and H.S. Harris (Indianapolis and Cambridge: Hackett, 1991), 149 [§93]: 'something becomes an other, and this other is itself a something, which, as such, then alters itself in the same way, and so without end'.

71. Nishitani, *Religion and Nothingness*, 176.

you be freed of birth and death'.[72] It is by negating both birth and death that Existence can be raised to such a height that it transcends nihility.

Let us take up an example that Nishitani gives, in order to understand what he means by a 'non-substantial understanding of being'. If one asks 'What is a fire?', then one looks for the *eidos* of the fire under the condition that 'fire here displays itself and displays itself to us'.[73] Substance is presented in terms of *logos*, as something to be explained in a logical and theoretical way through categories as it is in Aristotle. However, if we say that (a) 'fire doesn't burn fire', then (b) it is non-burning, and therefore it is fire:

> Substance denotes the self-identity of fire that is recognized in its *energeia* [...] on the contrary, the assertion that fire doesn't burn fire indicates the fact of the fire's 'non-burning', an action of non-action.[74]

To clarify this paradox, we can present it in this way: if fire, according to the substantialist thinking, is defined as that which burns, the fact that fire does not burn fire is the first step in moving away from the self-identity of fire as a substance, towards its *energeia* as such, towards another self-identity which is, for fire in itself, its 'home ground'.[75] In doing so, since fire is not seen as something burning—which is its essence from the substantialist point of view—it reclaims its

---

72.   Ibid., 178.

73.   Ibid., 113.

74.   Ibid., 116

75.   Ibid.

'true' identity, and therefore is fire.[76] That 'non-burning' is 'an action of non-action' means that the action of fire manifests itself in the privation of its substantialist form, and that therefore fire finds its definition in another ground. This constant negation neither ends at some definite point, nor does it become an infinite regression; rather, it seeks to maintain itself in a state in which substantial thinking is prevented from appropriating it:

> In contrast to the notion of substance which comprehends the selfness of fire in its fire-nature (and thus as being), the true selfness of fire is its non-fire-nature. The selfness of fire lies in non-combustion. Of course, this non-combustion is not some thing apart from combustion: fire is non-combustive in its very act of combustion. It does not burn itself. To withdraw the non-combustion of fire from the discussion is to make combustion in truth unthinkable.[77]

Nishitani wants to find the 'home ground' of fire, which consists neither in its actuality as fire nor in its potentiality to burn, but rather in its own ground which is defined by 'non-combustion', 'not burning itself'. However, this doesn't come from scientific observation, but rather from the privation of 'emptiness' in the Buddhist sense. Here we can see that Nishitani is attempting to carry out a similar task to that which Mou Zongsan attempted, although the latter used Kantian terminology whereas the former was strongly influenced by Heidegger and his language. Both of them argue that, whereas theoretical reason cannot enter the realm of the

---

76. Here Nishitani also refers to Heidegger's *a-letheia* (*Un-verborgenheit*).

77. Ibid., 117

noumenon, 'intellectual intuition' is able to arrive at theoretical reason through a self-negation. One can only enter the 'good infinity' that defines *Existenz* with a different kind of thinking:

> Infinity, as a reality, is cut off from the prehension of reason. No sooner do we try to grasp it in the dimension of reason than it turns forthwith into something conceptual.[78]

Is this logic enough to allow us to develop an East Asian thinking of modern science and technology, though? It is not possible to construct technics in the realm of the noumenon (although this word is only employed by Mou Zongsan, not Nishitani, to characterise such an infinity)—with the sole exception, perhaps, of the demiurge in Plato's *Timaeus*. Unlike the will for the second coming of Christ, which functions as the historical progress of the spirit, in the East Asian culture that Nishitani describes, the Will as Non-Will is detached from all historical happenings. Likewise, the noumenal thinking that Mou characterises seems to be another kind of historical consciousness, since it is not a matter of waiting for any event, but rather is subsumed under an order that is already ahead of history—it is a cosmological consciousness.

§24. OVERCOMING MODERNITY

Towards the end of *Religion and Nothingness*, Nishitani posed a question that he was not able to answer, though he attempted to do so throughout almost his entire career:

> [H]istorical consciousness has since seen remarkable devel-
> opments in the West. Particularly in the modern age, human

---

78.   Ibid., 177.

life itself gradually came to take shape through the historical self-consciousness of man. But what is involved in such a
development?[79]

Nishitani identifies the difference between West and East with
the fact that the former has developed a stronger notion of
historical consciousness. Understanding why such a historical
consciousness did not develop in the East is key to the relation
between technics and time in the East. In fact, this question
already troubled Nishitani during his early career, and was to
play an important role in his political philosophy. It is necessary
to examine this point, since it demonstrates both the necessity
and the danger of historical consciousness.

Between 1940 and 1945, Nishitani was deeply engaged in
the project of 'overcoming modernity' with his Kyoto School
philosopher colleagues, including Kōsaka Masaaki and Kōyama
Iwao (they were all students of Nishida Kitarō [1870–1945] and
Tanabe Hajime [1885–1962] at Kyoto Imperial University), as
well as the historian Suzuki Shigetaka (1907–1988). Nishitani's
ideas from this period are recorded in writings which include
the discussions organised by the literary magazine *Chuōkorōn*
in 1941–42 (the first being the famous 'The Standpoint of
World History and Japan'), the monograph *View of the World
and View of the Nation* (1941), and the essays 'My View of
"Overcoming Modernity"' (1942) and 'The Philosophy of World
History' ('Sekaishi no tetsugaku', 1944). Many scholars and
historians have already elaborated on the question of nationalism and imperialism embedded in these discussions and texts,[80]

---

79.   Ibid., 206.

80.   See, for example, N. Sakai, *Translation and Subjectivity: On 'Japan' and
Cultural Nationalism* (Minneapolis: University of Minnesota Press, 1997); and

and I will not repeat their arguments here, but will focus instead on the question of world history and historical consciousness.

Nishitani's project of overcoming modernity consists in the desire to return to Japanese culture and to transcend the Western culture and technology that had been imposed upon Japanese society since the arrival of the 'Black Ships' (Western vessels) in the sixteenth and nineteenth centuries. According to Nishitani, Western culture and technology had created a huge gap between tradition and modern life, and the Buddhism and Confucianism that once grounded Japanese society were no longer able effectively to engage with political and cultural life. This observation of Nishitani's evidently resonated with that of his colleagues, as well as that of Chinese thinkers of the same period. The New Confucians, for example, proposed to develop Chinese philosophy so that it could integrate Western rationality as one of its possibilities. We saw in Part 1 how, in Mou Zongsan, the question of *xin* (心, heart) or *liangzhi* (良知, 'conscience') formulated by Wang Yangming was evoked in order to descend from noumenal experience to the knowledge of phenomena. The *shin* (the Japanese equivalent of *xin*) is equally important for Nishitani in his rearticulation of the question of consciousness, and hence of historical consciousness; however, in Nishitani it also opens onto something else: absolute nothingness. In fact, it is immediately evident that, although they followed similar intellectual trajectories, the Chinese and Japanese thinkers developed two different responses to modernisation.

We should say a few words about Nishitani's teacher Nishida Kitarō here, since it was Nishida who developed the

C. Goto-Jones, *Re-Politicising the Kyoto School as Philosophy* (London and New York: Routledge, 2008).

concept of absolute nothingness. But Nishida also engaged with Wang Yangming's teaching on the unity of acting and knowing (知行合一), performing a combined reading of Wang YangMing with Fichte's concept of *Tathandlung* as well as William James's notion of 'pure experience'.[81] Fichte uses *Tathandlung* to describe a self-positing (*selbst-setzend*) beginning, which is not conditioned by something else—the *Unbedingte*, which firstly means the absolute or unconditional, but also means that which cannot taken as a thing (*Ding*). Nishida argues that it is not the knowing subject that comprehends reality; rather, the reality thus experienced constitutes the knowing subject. Nishida defines pure experience as the 'direct seeing of the facts just as they are', where 'direct seeing' is the Japanese translation of the German *Anschauung*.[82] Here the subject is not the absolute, but rather pure experience, which overcomes the isolationism of Fichte's *Ich*, and resonates with Mou Zongsan's characterisation of Wang Yangming's *liangzhi* as intellectual intuition. Nishida later further explored the condition of possibility of such an intuition, moving from Wang Yangming to the teachings of the Zen masters Dōgen

---

81. Kosaka Kanitsugu, 'Nishida Kitarō und Wang Yangming—ein Prototypus der Anschauung der Wirklichkeit in Ostasien', in Hsaki Hashi (ed.), *Denkdisziplinen von Ost und West* (Nordhausen: Raugott Bautz Verlag, 2015), 123–58. Feenberg assimilates this 'action-intuition' (行為的直觀) to what Heidegger calls circumspection (*Umsicht*), but is mistaken in doing so since, as we will see below, it is in fact related to intellectual intuition. See A. Feenberg, 'The Problem of Modernity in the Philosophy of Nishida', in J. Heisig and J. Maraldo (eds), *Rude Awakenings: Zen, the Kyoto School and the Question of Nationalism* (Hawaii: University of Hawaii, 1995), 151–73.

82. J. W. Heisig, *Philosophers of Nothingness An Essay on the Kyoto School* (Honolulu: University of Hawaii Press, 2001), 43; see Part 1 (§18.1), where we distinguish between Mou's concept of intellectual intuition and that of Fichte and Schelling.

and Shinran (1173–1262)[83] on nothingness. Nishida states that, if the West has considered being as the ground of reality, the East has taken nothingness as its ground[84]—*nothingness*, which 'does not itself come to be or pass away', is opposed to the world of being, and is *absolute* in the sense that it is 'beyond encompassing by any phenomenon, individual, event or relationship in the world'.[85] This absolute nothingness stands as the highest principle of reality, which Nishida calls the 'universal of the universals', since it relativises all other universal thoughts.[86] This 'nothingness' is not easy to understand. Firstly, it is paradoxical to pose the question 'What is the nothing?' because this immediately turns it into a question of being. Secondly, neither can one say that it is not real; in fact, according to Nishida, it has a place (*basho*, 場所) where being and/or nothing manifests,[87] though this may suggest that it *exists*.[88]

83.　F. Girard, 'Le moi dans le bouddhisme Japonais', *Ebisu* 6 (1994), 97–124: 98.

84.　Heisig, *Philosophers of Nothingness*, 61.

85.　Ibid., 62.

86.　Ibid., 63.

87.　Nishida's concept of *basho* consists in the idea that when two things are in relation, the relation always presuppose a place: if we consider the relation between A and not-A, there must be a place for such relation to take place. For a more detailed analysis of Nishida's concept of the space, see Augustin Berque, *Écoumène. Introduction à l'étude des milieux humains* (Paris: Belin, 2000), 53, 140.

88.　The question of the nothing is notably addressed by Heidegger in his 1929 Freiburg inauguration speech 'What is Metaphysics?', in W. McNeil (ed.), *Pathmarks* (Cambridge: Cambridge University Press, 1998), 82–96. Heidegger attempts to show that anxiety makes manifest the nothing, since in anxiety beings slip away: 'In anxiety beings as a whole become superfluous. In what sense does this happen? Beings are not annihilated by anxiety, so that nothing is left. How could they be, when anxiety finds itself precisely in utter impotence with regard to beings as a whole? Rather, the nothing makes itself known with beings and in beings expressly as a slipping away of the whole'. It is very possible that Heidegger is using the logic of privation here, in which respect his thinking is similar to that of the Buddhist thinkers. When Heidegger says that

Andrew Feenberg summarises the concept as 'experience as a field of immediate subject-object unity underlying culture, action and knowledge, and making them possible as objectification of the prior unity'.[89] For Nishida, absolute nothingness is the 'spiritual essence' that needs to be added to the 'materialism of the West' in order to bring about a correct order.[90] The concept of absolute nothingness is further developed by Nishida's colleague Tanabe, following a Hegelian logic, into a political and historical concept that provides a 'unifying *telos* to history'.[91] Nishitani's work takes this approach further: for him, absolute nothingness is no longer theoretical and individual; he believes that it can be concretely applicable to nations. How can absolute nothingness be understood in this way? In his book on nihilism, Nishitani states:

> I am convinced that the problem of nihilism lies at the root of the mutual aversion of religion and science. And it was this that gave my philosophical engagement its starting point, from which it grew larger and larger until it came to envelop nearly everything [...] the fundamental problem of my life [...] has always been, to put it simply, the overcoming of nihilism through nihilism.[92]

'Da-sein means: being held out into the nothing' (91), he means that one goes beyond beings as a whole, toward transcendence. This lecture was attacked by Rudolf Carnap in an article entitled 'The Overcoming of Metaphysics Through Logical Analysis of Language'. However, Wittgenstein, in his notes, left a paragraph showing his sympathy with Heidegger. Both Carnap's and Wittgenstein's texts can be found in M. Murray (ed.), *Heidegger and Modern Philosophy: Critical Essays* (New Haven, London: Yale University Press, 1978).

89. Heisig and Maraldo (eds.), *Rude Awakenings*, 160.

90. Heisig, *Philosophers of Nothingness*, 104.

91. Ibid., 121.

92. Ibid., 215.

Nishitani applies this same quasi-Nietzschean logic in proposing to 'overcome nationalism through nationalism'. He imagines a nationalism that is different from that of the modern nation-state—indeed, one that consists in the negation of the latter. Nishitani sees the modern state as a form of 'substrating' which aims to lay bare the ground of common unity; the negation takes place when individual freedom consciously appropriates control of—and thereby finally subjectivises—the state.[93] This is another form of nationalism which leads neither to an absolutism of the state, nor to a liberalism that would separate the individual from the state. To transcend the limits of the modern nation state, for Nishitani, it is by no means sufficient just to return to traditional Japanese values; one must construct a Japanese nation from the standpoint of world history. In doing so, Nishitani suggests, one can produce a 'leap from the subjectivity of a national ego to that of a national non-ego'.[94] The nation now takes the form of a subject, the unity of which consists in the will of all free individuals.

The Kyoto School project resonates with that of nineteenth-century Idealism—which is no coincidence, given that Nishitani began his academic career as a reader of Schelling, translating the latter's *Treatise on the Essence of Human Freedom* and *Religion and Philosophy* into Japanese; and that Nishida and Tanabe were very much interested in Hegel. However, the Kyoto school also set itself the task of transcending the Idealist project, as stated in Nishitani's *Philosophy of World History*:

The world of today demands that a new relation between

---

93.   Ibid., 197.

94.   Ibid.

world-historical research and the philosophy of world-history is thought differently from Hegel's philosophy of world-history, as well as Ranke's. And, furthermore, it demands that Hegel's reason of the state and rational idealism, and Ranke's *moralische Energie* and historical idealism—by transcending even the standpoints of even such great men—are reconsidered in a still more fundamental way.[95]

We can understand what Nishitani means here by 'world-historical research' and 'philosophy of world-history' by briefly reflecting on the two camps of the debate on historicism in Germany during the period from 1880 to 1930. On the one hand, there was the dominant model for academic research among neo-Kantians such as Wilhelm Windelband and his student Heinrich Rickert; on the other hand there were those whose concept of history was attacked for a tendency toward 'relativism', such as Friedrich Meinecke's 'vitalist' view and Dilthey's *Weltanschauungslehre*.[96] This debate came to an end after Heidegger's *Destruktion* of ontology in *Being and Time*. The Kyoto school sought to overcome Hegel's rational idealist view of history as the realisation of the spirit—a notion not so far from Leibnizian theodicy, given that Hegel says that history is 'the justification of the ways of god'.[97] But they also wished to overcome Leopold von Ranke's description of history as a set of unique and singular events driven by 'moral energy'.

---

95.   Cited by Christian Uhl, 'What Was the "Japanese Philosophy of History"? An Inquiry Into the Dynamics of the "World-Historical Standpoint" of the Kyoto School', in C.S. Goto-Jones (ed.), *Re-Politicizing the Kyōto School as Philosophy* (London: Routledge, 2008), 112–34: 125; Keiji Nishitani, *Chosakushū*, IV (Tokyo: Sōbunsha, 1987/1988), 252.

96.   See Bambach, *Heidegger, Dilthey and the Crisis of Historicism*.

97.   See P. Chételat , 'Hegel's Philosophy of World History as Theodicy. On Evil and Freedom', in W. Dudley (ed.), *Hegel and History* (New York: SUNY Press, 2009), 215–30.

In short, the thought of the Kyoto School was strongly influenced by German philosophy and, whether consciously or unconsciously, they took up the philosophical task of Germany to reformulate a philosophy of the world-history without a Christian aim, as if it were now Japan's responsibility—as Suzuki overtly declared during the *Chuōkorōn* meeting, in a call that resonates with Heidegger's 1933 inaugural speech as the Rector of Freiburg University:

> According to Hegel, it was the Roman and German people who carried the destiny of world-history on their backs, but today it is Japan that has become conscious of such a world-historical destiny [...] The reason why Japan possesses leadership in East Asia lies with the fact that Japan is conscious of its world-historical destiny, which actually *is* this consciousness. This destiny is not saddled on Japan objectively, but Japan makes it subjectively conscious to itself.[98]

This task consists in overcoming the limits of European culture, as actualised in present forms such as the nation state, capitalism, individualism, and imperialism; in other words, the impasse of European modernity. According to the Kyoto School, it is up to the Japanese nation to overcome this legacy by creating a new world history through a nationalism and imperialism proper to it[99]—and the only way to realise this whole project is a 'total war' (*sōryokusen*, a translation

---

98. Uhl, 'What Was the "Japanese Philosophy of History"?', 120; cited from 'Tōa kyōeiken no rinrisei to rekishisei', *Chūōkōron* (April 1942): 120–27: 127.

99. T. Kimoto, *The Standpoint of World History and Imperial Japan*, PhD Thesis, Cornell University, 2010, 153–5.

from the German *totaler Krieg*).[100] This total war is presented as a purification through which new subjectivities will arise from the lost Japanese spirit and realise absolute nothingness as the ground for a 'universal world history' in which many 'specific world histories' can 'exist harmoniously and interpenetratingly'.[101] This 'total war' is thus an 'accelerationist' strategy *par excellence* that seeks to intensify the conflicts between states and individuals in order to transcend the world as objective totality. War, for the Kyoto school philosophers, is the force that defines history and therefore world history.[102] We might well say that the Idealists' concept of *strife* (*Streit*) is reincarnated in the concept of war here. Idealists such as Schelling, Hölderlin, Hegel, and the early Romantics found in Greek tragedy a literary form which expresses such a strife: tragedy is based on the necessity of fate, and the tragic hero affirms the necessity of suffering as the realisation of his freedom.[103] In the Japanese version, though, tragedy finds its realisation in a vision of 'world history as purgatory'.[104] To the eyes of the Kyoto School, the Sino-Japanese war had nothing to do with imperialism, but happened because it was the moral obligation of Japan to save China.[105] The realisation of the Greater East Asia Co-Prosperity Sphere is one part of the

100. Ibid., 148.

101. Ibid., 149.

102. Uhl, 'What Was the "Japanese Philosophy of History"?', 115.

103. See D. J. Schmidt, *On Germans and Other Greeks: Tragedy and Ethical Life* (Indianapolis: Indiana University Press, 2001).

104. Kimoto, *The Standpoint of World History*, 145.

105. Note that Suzuki also claimed, during the Chūōkōron meeting, that 'we must conclude that morality did exist in China, but no moral energy'; see Uhl, 'What Was the "Japanese Philosophy of History"?', 123; cited from *Chūōkōron* (April 1942), 129.

new history that Japan is 'obliged' to realise for the benefit of East Asia. The conception of this 'just war' is given in Kosaka Masaaki's concluding statement of the first roundtable section for the magazine *Chūōkorōn*:

> When man becomes indignant, his indignation is total. He is indignant in both mind and body. This is the case with war: both heaven and earth become indignant. In this way, the soul of humanity comes to be purified. This is why it is war that determines the crucial turning points in world history. Hence world history is purgatory.[106]

Looking back at this fanaticism, one can find in Nishitani's philosophy a justification for a kind of racism and nationalism, which are seen as the 'means' towards their own negation—a negation that moves towards absolute nothingness and the political project of a world history radically different from that defined solely by Western modernity.

The different intellectual milieus of China and Japan yielded different interpretations of modernity, then. It might be said that the Japanese intellectuals experienced a deeper problem of time and history, and that it was the question of time qua history that they sought to overcome. Chinese intellectuals such as Mou Zongsan, on the other hand, were puzzled by the question of why modern science and technology had not arisen in China, and concluded that this probably owed largely to China's long intellectual history, which has a totally different philosophical temperament from that of the West.

---

106. Cited by Kimoto, *The Standpoint of World History*, 145; from Kōsaka et al., 'Sekaishiteki tachiba to Nihon', 192. Kimoto also notes that this concept comes from Nishida.

Therefore, as we have seen, for Mou Zongsan the strategy was to show that, while retaining its traditions and moral teachings, it would nevertheless be possible for China to produce the same modernity as in the West, and to bypass the crisis in Europe.[107] Nishitani (and the other members of the Kyoto School), on the other hand, tried to show that the only way to transcend the nihilism of modernity was to reconstitute world history on the ground of absolute nothingness.

Like Nishida, Mou Zongsan also engaged with Wang Yangming and later with Buddhism, yet he arrived at a totally different reading. His later work *On the Highest Good* (圓善論, 1985) is based on a new interpretation of Tian Tai Buddhism, which, as he admitted in *Nineteen Lectures on Chinese Philosophy* (1983), he had not engaged with in his *Phenomenon and Thing-in-itself*.[108] In *On the Highest Good* he found that the 'perfect teaching [圓教]'[109] in Tian Tai Buddhism is more advanced than the teaching of 'one mind opens two doors' which had allowed him to resolve the 'ontological difference' between phenomenon and noumenon as well as the Kantian

---

107. Unlike the Kyoto School philosophers, Mou rarely touched upon the question of modernity. Some commentators, such as Stephan Schmidt, have proposed that there is a hidden agenda in Mou's teaching, for example, a 'declaration of intellectual independence of Confucian philosophy from Confucian institutions' (S. Schmidt, 'Mou Zongsan, Hegel and Kant: the Quest for Confucian Modernity', *Philosophy East and West* 6:2 [April 2011], 260–302: 276). This is not entirely convincing, however. Mou was one of the founders of the New Asia College, which is now part of the Chinese University of Hong Kong, and served as a professor in the universities throughout the rest of his career.

108. Tomomi Asakura, 'On Buddhistic Ontology: A Comparative Study of Mou Zongsan and Kyoto School Philosophy', *Philosophy East and West* 61:4 (2011), 647–78: 649.

109. 'Perfect teaching' for Mou means a teaching that cannot be achieved through linguistic description, but which must necessarily go beyond language, Mou Zongsan, *Nineteen Lectures*, 248.

antinomy of practical reason. In fact, in *Intellectual Intuition and Chinese Philosophy* (1971), when explaining intellectual intuition in Buddhism, Mou already hinted at the superior teaching of Tian Tai Buddhism over others,[110] a superiority that can be appreciated both through its critiques of the other teachings and through its own propositions. Firstly, Tian Tai Buddhism accused Hua Yuan Buddhism of cutting off the lower worlds (i.e. the worlds of animals, spectres, etc.) and abandoning them, while emphasising the purity of truth (緣理斷九).[111] This critique in fact resonates with Wang Yangming's critique of Buddhism, which holds that it aims only to transcend beings without taking care of them, meaning that Confucianism is superior to Buddhism as a form of social and political thought. The 'perfect teaching' of Tian Tai consists in the expression 'three thousands realms in a single instant of thought' (一念三千), which for Mou is a fuller expression of intellectual intuition than 'one mind opens two doors'.

As Tomomi Asakura has acutely pointed out, one can understand the difference between Mou and the Kyoto School by considering that Mou departs from a moral standpoint whereas the Kyoto School begins with a religious standpoint— in the philosophy of Tanabe, 'the attitude that sees the reality qua absolute contradiction and absolute self-disruption'.[112] Mou sought an 'internal transcendence (內在的超越)' within his 'non-attachment ontology [無執的存有論]', whereas Nishitani sought an overcoming which took its most radical form by achieving emptiness through war. What is at stake in both enterprises, however, is the problem of time, and of a history

---

110. Mou Zongsan, *Intellectual Intuition and Chinese Philosophy*, 211–15.

111. Ibid., 215; Asakura, 'On Buddhistic Ontology', 661.

112. Asakura, 'On Buddhistic Ontology', 666.

that has been totally conquered by an *axis of time* largely defined by European ontotheology and its completion in the realisation of modern technology. If the failure of both of these projects—though for different reasons, since the Kyoto school's decline owed largely to Japan's defeat at the end of the Second World War—has anything to tell us, it is that, in order to overcome modernity, it is necessary to go back to the question of time and to open up a pluralism which allows a new world history to emerge, but one which is subordinated neither to global capitalism and nationalism, nor to an absolute metaphysical ground. This new world history is only possible by undertaking a metaphysical and historical project, rather than simply claiming the end of modernity, the end of metaphysics, the return to 'nature'—or, even less credibly, the arrival of the multitude.

Further analysis of the historical consciousness at the centre of Nishitani's programme for overcoming modernity will enable us to answer the question of the lack of historical consciousness in East Asian culture. Firstly, let us bear in mind that Granet's and Jullien's observations on the question of time in China are no less applicable to Nishitani. During the 1970s, Nishitani held several discourses in different temples in Japan discussing modernisation and Buddhism, later published as a book entitled *On Buddhism*. As suggested above, we may suppose Nishitani to have been haunted by the question of historical consciousness; and indeed, at a certain point, he claims that the concept of the historical does not exist in East Asian culture. What he means by 'historical' is the awareness of situating oneself as a historical being, and anamnesis as reconstruction of historicity, *Geschichtlichkeit*. Retrospectively—at least from what he said during these lectures, and given his personal relation to Heidegger—Nishitani's concept

> I am sure that Buddhism falls short of such historical consciousness, at least to some extent. Generally speaking, something called 'historical' exists no less in China than in India and Japan. But I have the impression that in these countries there has been no trace of seeing the world as history in the true sense of the word [...] This way of thinking is somewhat different from an historical one, at least of the sort prevalent in the modern world.[113]

In claiming that Asia has not been capable of 'seeing the world as history in the true sense of the word', Nishitani means that in Eastern thinking there is a lack of elaboration of the temporalisation of the past, present, and future. Nishitani believes that the concept of history is intrinsic to Christianity.[114] In Christianity, the original sin and eschatology mark a beginning and an end, as well as the limit of waiting for a beginning of a new epoch, with the Second Coming of Christ. Christianity is historical also in a second sense, indicated by the human's progressive conception of himself in relation to God. For Nishitani, this historical consciousness genuinely arose during the Renaissance, and culminated during the Reformation. In the Renaissance, it was indicated by the consciousness that the world order is not entirely dependent on providence, and that the personal relation between God and man is cut across by the natural sciences;[115] and in the Reformation in the realisation that history is but a human product. In contrast, Nishitani

113. K. Nishitani, *On Buddhism*, tr. S. Yamamoto (New York: SUNY Press, 2006), 40.

114. Ibid., 56.

115. Nishitani, *Religion and Nothingness*, 89.

observes that, in Buddhism, there is a negativity in time which must be transcended, meaning that finitude in both its linear and cyclical form must be transcended in order to attain absolute emptiness. Therefore Buddhism is not able to open up the question of historical consciousness, and does not see the possibility of 'emergence' in every 'now'.[116] Nishitani continues:

> [T]he other aspect—namely, that it is historical and that being is time—is comparatively neglected. Or rather I should say, if the term 'neglect' is a bit of an exaggeration, it is not sufficiently developed. This is attributable to the fact that Buddhism places emphasis on the negative inherent in the contention that time is somewhat transient and that this is a world of suffering. Buddhism seems to have failed to grasp that the world of time is a field in which something new emerges without interruption.[117]

Nishitani employs a Heideggerian vocabulary here, using *Augenblick* to translate the 'now' which, for Heidegger, acts as a vertical cut into the flow of time.[118] Heidegger had used it to translate the Greek word *kairos*, which suggests a non-chronological time that can be presented as a rupture or jump. There is certainly a sense of rupture and jump in Zen Buddhism, namely in what is called *dun wu* (頓悟, 'insight'). A Buddhist becomes a master in the moment of *dun wu*, which happens like a flash over the sky—for example, it takes place at the moment when one sees a frog jumping into the

---

116. Nishitani, *On Buddhism*, 50.

117. Ibid., 49–50.

118. 'History and superhistory, time and eternity, cross over and intersect with one another. This point of intersection has been called the "now," "here," or "the point of contact." As you know, to use terminology prevalent in the West, it is often termed the "moment," (that is, *Augenblick* in German). The "now," while ex-isting in time, cuts time in a vertical way.' Ibid., 49.

pond, as described in the haiku of Matsuo Bashō (1644–1694), 'old pond—a frog jumps in, water's sound'.[119] But *dun wu* neither happens in every now, nor does it require a long goal-seeking process: it is one time for all, since *dun wu* is a radical transformation or elevation which opens a new realm of experience and a new way of thinking. It transcends time, and one may describe it with Nishitani's own term 'superhistory'. Nishitani's attitude on this question in relation to Heidegger differs from that of Mou Zongsan: As we saw, Mou claimed that Heidegger failed to understand that, although Dasein is finite, it can also transcend itself and enter the infinite through intellectual intuition.

Following Heidegger, Nishitani also makes a distinction between *Geschichte* and *Historie*. *Historie* means 'to talk about stories or to hand down legends' whereas *Geschichte* indicates that 'something happens, or that something novel that has never been before has arisen'.[120] Like Heidegger, he links *Geschichte* to the verb *geschehen*, meaning 'happening'. History as *Geschichte* is linked to the *Ereignis*, conditioned by a historical consciousness. In such a consciousness, past, present, and future as such become contemporaneous with each other. It therefore no longer involves a subject that looks back at the past as a series of historical events, but one that sees itself as a historical being, grounded in the hermeneutics of history.

One must be cautious here to avoid squaring Nishitani's interpretation of Buddhism with Mou Zongsan's—not to mention to avoid equating Buddhism with Chinese culture. However, we are at least justified in saying that, in both of these cultures,

119. Matsuo Bashō, *Bashō's Haiku*, tr. D.L. Barnhill (Albany, NY: State University of New York, 2004), 54.

120. Nishitani, *On Buddhism*, 74.

the notion of time was not only underdeveloped, but was also taken as something to be transcended. Such transcendence is within the capacities of intellectual intuition, which in Mou's work allows the subject access to the moral cosmology, nature, and the void.

Returning now to Stiegler's thesis of technics as originary question, we must also recognise, once again, that such a historical consciousness also hinges on a series of technological inventions such as copying and printing (especially when one considers the printing of Bibles during the Reformation)— meaning that, besides Christian eschatology, it is also technicity that allows the 'now' to occur as a vertical cut, as *Ereignis*. A dialogue can be set up between Nishitani and Stiegler precisely on the basis of Stiegler's critique of Heidegger on this point: Stiegler showed that Heidegger considered world history only as the possibility of Dasein, without recognising how exteri-orisation is necessary for the constitution of Dasein; meaning that the concept of world history in *Being and Time* remains a transcendental discourse.[121] Let us look at how Heidegger defines world history:

> With the existence of historical being-in-the-world, what is ready-to-hand and what is present-at-hand have already, in every case, been incorporated into the history of the world. Equipment and work—for instance, books—have their 'fates'; buildings and institutions have their history. [...] These beings within the world are historical as such, and their history does not

---

121.   B. Stiegler, *Technics and Time. 2: Disorientation* (Stanford, CA: Stanford University Press, 2009), 5.

signify something 'external' which merely accompanies the 'inner'
history of the 'soul'. We call such beings 'the world-historical'.[122]

The world-historical is what Stiegler calls 'tertiary retention'.[123]
Heidegger didn't disregard technics, of course; neither did
he fail to reflect that the world into which Dasein is thrown
functions as an 'already-there', which he presents as 'factic-
ity'. But he did not consider Dasein's temporalisation from the
perspective of technics, which is also the condition of such
temporalisation; rather, he gives the ultimate possibility to
Dasein in the realisation of being-towards-death. Stiegler's
critique is that the world-historical is 'not simply the result
of what falls behind the temporalising who in the form of
traces', but is the very constitution of the 'who in its proper
temporality'.[124] This historicity has to be retrieved through
the anamnesis of writing, or technics. As Stiegler further
shows in the third volume of *Technics and Time*, writing is the
'spatialisation of the time of consciousness past and passing
as *Weltgeschichtlichkeit*'.[125] Technical objects are the second
spatialisation of time as interval, and historicity is only possible
through anamnesis with the aid of mnenotechnics.

We can see a kind of contradiction here between Nishitani's absolute nothingness and Stiegler's positing of the
necessity of a world history in a technical form. World history
is fundamentally an anamnesis par excellence, while absolute

---

122.   Cited by Stiegler, *Technics and Time 1*, 237; from Heidegger, *Being and
Time*, 388 [translation modified].

123.   Stiegler, *Technics and Time 3*, 37.

124.   Cited by Stiegler, *Technics and Time 1*, 237; from Heidegger, *Being and
Time*, 310.

125.   Stiegler, *Technics and Time 3*, 56.

nothingness is the original ground freed from any relativity, a self-grounding absolute emptiness or void. This is not to perform a deconstruction of Nishitani's concept of historical consciousness, but rather to suggest that, behind historical consciousness, there lies the functioning of technological unconsciousness. In other words, Nishitani could by no means achieve his goal without a 'psychoanalysis' of the technological unconscious of East Asian culture. If it is true, as Stiegler claims, that there can be no historical consciousness without a technical support, then we may say that in China and in Japan, there must also be a kind of technological unconsciousness, in so far as there are historical writings, printings, and other techniques which were, in their time, as refined and sophisticated as any in the world. But then we will have to explain why this technical support—these writings and printing techniques, which existed in China in an equal if not more advanced state than in Europe in ancient times—did not give rise to such a historical consciousness in China. Certainly, we are not claiming that these technics in China and Japan have nothing to do with memory. Rather, we seek to show that a philosophical system that functions on the basis of intellectual intuition, and seeks insight into the noumenon, refuses to take this memory into account; and that what results is a division into two ontologies, namely a noumenal ontology and a phenomenal ontology, where the dominance of the former implies the subordination of the latter.

Now, this may seem to lead to a circular argument: (1) there is a lack of historical consciousness, because the question of time is not elaborated; (2) since the question of time is not elaborated, the relation between technics and time has never been a question; (3) since the relation between technics and time is not a question, a historical consciousness as anamnesis does not arise. But it is at this point that the question of geometry, time, and anamnesis discussed above returns, and returns us to the question of the relation between *Qi* and *Dao*.

Qi may be described, in Stiegler's terms, as a 'retentional object', since as a technical object it retains traces, or memories. But in China it is ontologically a-temporal, a-historical, since it accords with *Dao* and expresses *Dao*—and in order to be 'excellent', nothing is more important than to accord with *Dao*. The *Dao* here is a cosmological and a moral one; *Qi* is part of the cosmology, but it is governed by a principle that is not defined by and in itself, but by its relation to other beings, both human and non-human. In Part 1 (§14–15), we discussed the line of thought on the relation between *Dao* and *Qi* advocated by Zhang Xuecheng and Wei Yuan. Zhang Xuecheng acutely pointed out that one should consider the relation between *Qi* and *Dao* as a historical-temporal one. For him, the six classics are historical artefacts, and therefore belong to *Qi*. This de-absolutises the relation between *Dao* and *Qi*, while at the same time leads to a re-inscription of it, simply because according to this account, *Qi* carries the *Dao* of the epoch. Wei Yuan's inverse attempt to inscribe *Dao* in *Qi*, proposed after the two Opium Wars, was an attempt to secure a technological consciousness. However, such a consciousness is unable to hold sway, and is flushed away immediately by the 'Cartesian separation' of Chinese thought as mind from Western technology as mere instrument. We may well see both cases as efforts to mobilize the ontological *Qi-Dao* relation in order to produce a new *episteme*.[126]

---

126.   For the use of the term *episteme*, please refer to §2.

§25. ANAMNESIS OF THE POSTMODERN

Forty years after the demise of the Kyoto School, the task of 'overcoming modernity' adopted another form in Europe: the 'postmodern', made famous by Lyotard. Indeed, the task with which Nishitani charged himself—namely, to overcome European culture and technology with absolute nothingness—was to find certain resonances in Lyotard's formulation of the postmodern. I refer here specifically to Lyotard's text '*Logos and Techne*, or Telegraphy', included in the 1998 collection translated into English as *The Inhuman: Reflections on Time*. Lyotard first gave this paper at a 1986 seminar of IRCAM (*Institut de Recherche et Coordination Acoustique/Musique*) at the Centre Georges Pompidou organised by Bernard Stiegler, who was at the time writing his masters' thesis under the supervision of Lyotard. The text specifically addressed the question of anamnesis and technics, a theme that was to become central to Stiegler's philosophy.

The main thesis of this seminar consists in the following: the relations between matter and time, according to Lyotard, can be grasped in three different temporal syntheses: habit, remembrance, and anamnesis. Habit is a synthesis that expresses itself bodily. Remembrance seeks a narrative with an origin, or a beginning. Anamnesis, for Lyotard, means something rather different, and must be carefully distinguished from remembrance. This distinction has its source in Freud, especially his 1914 essay 'Remembering, Repeating, and Working-Through [*Erinnern, Wiederholen und Durcharbeiten*]'. In this essay Freud tried to show that there are two techniques of analysis: hypnosis, which helps the patient to reconstruct the unconscious content through a simple form of remembering (simple in the sense that the patient is removed from the present, and what matters is the earlier situation); and a second scenario in which 'no memory can as a rule be recovered'.[127] This second situation occurs, for example, with certain childhood experiences which we didn't understand at the time, but which somehow disclosed themselves subsequently. The most significant difference between the technique of remembrance in hypnosis and the technique of uncovering repetition is that, in the latter, the patient 'reproduces it not as a memory but as an action; he repeats it, without, of course, knowing that he is repeating it'.[128] The analyst's task in this case is to help the patient to uncover the source of resistance. However, as Freud identified, there are two difficulties here: the first is that the patient may refuse to recognise that there is a problem—that is to say, he or she may refuse to remember; the second is that novice analysts often found that, even after revealing this resistance to the patient, there was no change. It is at this point that Freud introduces the third term, *Durcharbeiten* or 'working-through':

> One must allow the patient time to become more conversant
> with this resistance with which he has now become acquainted,
> to *work through* it, to overcome it, by continuing, in defiance
> of it, the analytic work according to the fundamental rule
> of analysis.[129]

In '*Logos* and *Techne*, or Telegraphy', Lyotard makes reference to Stiegler's retentional model of memory (through spatialisation) by referring to three modes of memory—breaching

---

127. S. Freud, 'Remembering, Repeating, and Working-Through', *Standard Edition*, tr. J. Strachey (London: Hogarth, 24 vols., 1953–1974.), vol. 12, 147–56: 149.

128. Ibid., 150.

129. Ibid., 155.

(*frayage*), scanning (*balayage*) and passing (*passage*), corresponding respectively to habit, remembrance, and anamnesis—identifying Freud's *Durcharbeiten* with the third type of synthesis of time, anamnesis. Lyotard's reading of *Durcharbeiten*, however, is quite different from Freud's.[130] For Lyotard, this anamnesis has two different senses, the nuances of which must be carefully distinguished. The first sense of *Durcharbeiten* takes the form of free association: as Lyotard says, 'passing' takes more energy than scanning and breaching, precisely because it has no preestablished rules.[131] This sense is taken up on another occasion, in *The Postmodern Explained*, where Lyotard understands avant-gardism as a movement that bears great responsibility for the presuppositions implied in modernity. The work of the modern painters, from Manet to Duchamp to Barnett Newman, can, he suggests, be understood in terms of an anamnesis, in the sense of psychoanalytic therapeutics:

> Just as the patient tries to elaborate his present trouble by freely associating some apparently inconsistent elements with some past situation—allowing them to uncover hidden meanings in their lives and their behaviour—in the same way we can think of the work of Cézanne, Picasso, Delaunay, Kandinsky, Klee, Mondrian, Malevich, and finally Duchamp as a working through (*Durcharbeiten*) performed by modernity on its own meaning.[132]

---

130. In Dominique Scarfone's article 'À quoi œuvre l'analyse?', *Libres cahiers pour la psychanalyse* 9 (2004), 109–23, the author argues that, for Freud, the *Durcharbeiten* is a task that falls to the patient, and the analyst can only wait and allow things to occur; for Lyotard the contrary is the case: it is the 'third ear' of the analyst that allows the passage of the signifier.

131. Lyotard, '*Logos* and *Techne*, or Telegraphy', 57.

132. J.-F. Lyotard, *The Postmodern Explained: Correspondence, 1982–1985*, tr. D. Barry (Sydney: Power Publications, 1993), 79–80 (translation modified).

For Lyotard, these artists don't represent a rupture with the modern, but rather an anamnesis of the modern. Hence they are representatives of a postmodern art which liberates itself from rules and responsibility, and passes beyond the rules of inscription, through anamnesis. But yet more intriguing, albeit also somewhat puzzling, is Lyotard's demand for something which is not inscribed and hence cannot be limited by the rules of writing—an origin that is not something remembered, indeed, a memory that is not inscribed, and yet cannot be forgotten—as exemplified by Freud's notion of the experience of childhood as something that is not remembered but which must nonetheless be worked through. Christopher Fynsk proposes to emphasise the role of infancy in Lyotard's concept of anamnesis in relation to this point, noting that Lyotard 'understood himself to be writing *from* an infancy and *to* an infancy'.[133] In the section on anamnesis in '*Logos* and *Techne*, or Telegraphy', in a passage crucial for our inquiry here, Lyotard dramatically introduces an example from Dōgen in order to explain what he means by 'passing' or anamnesis. In this use of Dōgen we can observe the different nuances that mark out 'passing' from anamnesis as *Durcharbeiten*. Fynsk writes:

> I believe that the appeal to Dōgen, here, is not merely an instance
> of exoticism, however effective it might also be on that score. It
> is rather an implicit acknowledgment that what [Lyotard] seeks
> to think does not surrender to the concept or to any theoretical
> exposition—that if there is a passage from infancy to thought,
> it is not established by the concept.[134]

---

133. C. Fynsk, 'Lyotard's Infancy', *Yale French Studies* 99, *Jean-Francois Lyotard: Time and Judgment* (2001), 48.

134. Ibid., 55.

I would take this reference to Dōgen more seriously than Fynsk does. Indeed, the reference to Dōgen is not limited to this one occasion in Lyotard's writings, but recurs in various notes and interviews. What Lyotard was thinking here was much more intriguing and more uncanny than Fynsk suggests: it is none other than the logic that Nishitani used to avoid reducing being to its essence, as in his example about fire. I call this logic the *negation of logos*—although the word 'negation' is perhaps not entirely correct, since the negation here is neither a total negation nor a partial privation (e.g. part, intensity). We could clarify this distinction between privation and negation by paraphrasing the peculiar example Heidegger uses to clarify the difference as understood by the Greeks: when I am asked if I have time for skiing, I reply, 'no, I don't have time'. In fact, I do have time, but I don't have time for *you*.[135] Here being is not negated by taking a reverse direction, but rather subject to privation in such a way that it is taken out of its usual context (as in 'fire doesn't burn fire'). This logic is exemplified in the movement from modern to postmodern. The postmodern is the self-negation of the modern. It is not that, at a certain moment of modernity, something happened, and at that point the postmodern arrived; it is rather that, at a certain moment of its development, the logic of modernity turned against itself and transplanted itself into another context.[136] The reference to Dōgen, I believe, seeks to demonstrate the same logic, no

135. M. Heidegger, *Zollikon Seminars: Protocols, Conversations, Letters*, ed. M. Boss (Evanston, IL: Northwestern University Press, 2001), 46–47. Heidegger writes: 'It took Greek thinkers two hundred years to discover the idea of privation. Only Plato discovered this negation as privation and discussed it in his dialogue *The Sophist*.'

136. This negation, which emerges from an internal development, is the logic presented by Lyotard in his Introduction to *Les Immatériaux*. See J.-F. Lyotard, *Deuxième état des immatériaux*, March 1984 (Archive du Centre Pompidou).

longer limited to the case of modernity but applied to *logos* as such. I believe that here Lyotard poses his ultimate question on technics, even if it remains shrouded in ambiguity—namely, he attempts to compare what he means by anamnesis with what Dōgen, in *Shōbōgenzō*, the classic of Zen Buddhism, calls a 'clear mirror'. I will quote Lyotard's comment at length:

> It makes sense to try to recall something (let's call it something) which has not been inscribed if the inscription of this something *broke* the support of the writing or the memory. I'm borrowing this metaphor of the mirror from one of the treatises of Dōgen's *Shōbōgenzō*, the *Zenki*: there can be a presence that the mirror cannot reflect, but that breaks it into smithereens. A foreigner or a Chinese can come before the mirror and their image appear in it. But if what Dōgen calls 'a clear mirror' faces the mirror, then 'everything will break to smithereens'. And Dōgen goes on to make this clear: 'Do not imagine that is first the time in which the breaking has not yet happened, nor that there is then the time in which everything breaks. There is just the breaking.' So there is a breaking presence which is never inscribed nor memorable. It does not appear. It is not a forgotten inscription, it doesn't have its place and time on the support of inscriptions, in the reflecting mirror. It remains unknown to the breachings and scannings.[137]

This passage is undoubtedly the most puzzling part of Lyotard's intervention. The mirror and the clear mirror certainly have no end of metaphorical connotations. And yet, as Fynsk points out, it is very difficult for us to analyse this statement—to consider a dialogue between a twentieth-century French

---

137. Lyotard, '*Logos*', 55.

philosopher and a thirteenth-century Japanese monk—without falling into some kind of exoticism.

Taking up Dōgen further, the clear mirror stands for the mind (or intellectual intuition) before which the phenomenon is dismantled. The clear mirror presents something almost opposite to any conceptualisation of substance, since it is emptiness. Firstly, the clear mirror negates substance or essence (*ousia*) as *eidos*. This may remind us of Nishitani's perplexing logic of self-identification: fire doesn't burn fire, therefore it is fire. The phenomenal experience presents itself as such out of the actualization of the mind, since a normal person clings to substantializing it. The clear mirror is another type of mind, one able to operate a privation of this substantialist inclination; to this mind the world appears in constant change, without any persistence. There was no event that broke the clear mirror and marked the beginning. Before a clear mirror, there is only constant breaking, which destroys the concept of the self (the self cannot be mirrored at all). A person who doesn't have a mind like the clear mirror can see herself, since she still has *upādāna* (clinging, grasping, attachment), which only sees phenomena since it can only navigate through forms. In contrast, a clear mirror sees everything broken, since in-itself it is empty. Lyotard continues:

> I am not sure that the West—the philosophical West—has succeeded in thinking this, by the very fact of its technological vocation. Plato, perhaps, when he tries to think *agathon* beyond essence. Freud perhaps when he tries to think primary repression. But both always threatening to fall back into the *technologos*. Because they try to find 'the word that gets rid', as

Dōgen writes. And even the late Heidegger is perhaps missing the violence of the breaking.[138]

It is not clear how much Lyotard knew of the history of the 'clear mirror'—a story that is famous in Zen Buddhism, though it is considered to be apocryphal. According to the tale that is told, the Fifth Patriarch of Zen Buddhism Daman Hongren wanted to find a successor, and his student Shenxiu (606–706) was considered a strong candidate. However, Hongren had doubts and wanted to find a more suitable person, and, in order to select one, asked his students to write a poem to explain what the mind is. Shenxiu wrote the following on the wall:

身是菩提樹，   The body is the Bodhi tree,
心為明鏡台。   The heart is like the clear mirror.
時時勤拂拭，   One should often clean it,
勿使惹塵埃。   And let no dust alight.

Huineng (638–713), an obscure figure in the temple, responded with another poem. In fact, Huineng couldn't read or write, so he had to ask someone else to do it for him. (This is one of the characteristics of the practice of Zen Buddhism, where literacy is not considered to be an important virtue.) The poem won the approbation of Hongren, and Huineng became the Sixth Patriarch of Zen Buddhism. The clear mirror is the mind that Zen Buddhism seeks to achieve:

菩提本非樹，   Bodhi is not tree,
明鏡亦非台，   Clear mirror is not mirror.
本來無一物，   There is not a single thing,
何處惹塵埃。   How can it gather dust?

---

138. Ibid, 55.

Lyotard, though, transforms the clear mirror into a question of *writing*, and thus also a question of *logos*. Here we come across another meaning of substance: the support, or *hypokeimenon*. The question is: Can being be, without being carried by a *hypokeimenon*? Or, as Lyotard asked in the first text reproduced in *The Inhuman*, 'Can Thought Go On Without a Body?' Is *logos* able to facilitate an anamnesis that is not inscribed by it? In other words, can *logos*, and here *techno-logos*, rather than determining the anamnesis, allow it to arrive in a non-deterministic way? Lyotard thus hopes to *overcome logos through logos*, in the same way that Nietzsche and Nishitani sought to *overcome nihilism through nihilism*. Another similar passage in the teaching of Dōgen demonstrates this logic: the Zen master teaches '*Think of not-thinking. How do you think of not-thinking? Non-thinking. This is the essential art of zazen*' (*Zazen* or *tso-ch'an* literally means 'sitting Zen' and is a technique of meditation).[139] The opposition that Dōgen creates here is that between thinking and not-thinking. This is a pure negation, since thinking cannot be not-thinking, and not-thinking cannot be thinking. But for Dōgen, between thinking (*shiryō*) and not-thinking (*fushiryō*), there is a third way, which is *non*-thinking (*hishiryō*), and which negates both thinking and not-thinking, through the privation of thinking. For Lyotard, this privation of the *logos* leads to a realm which is not inscribed and is not inscribable in the *logos*. Lyotard adopts this logic himself when, in a talk given at a colloquium on the occasion of the opening of an exhibition of the artist Bracha Lichtenberg Ettinger, later

---

139. C. Olsen, *Zen and the Art of Postmodern Philosophy: Two Paths of Liberation From the Representational Mode of Thinking* (Albany, NY: State University of New York Press, 2000), 68.

published as 'Anamnesis of the Visible', he describes her work with the phrase: '*I remember that I no longer remember*'.[140] We might say that this double-bind is the logic of anamnesis: Is the non-*logos* possible through the negation of *logos* within *logos*? In the last paragraph of '*Logos* and *Techne*, or Telegraphy', Lyotard raises the question that we cited in the Introduction:

> [I]s the passage possible, *will it be possible with, or allowed by,* the new mode of inscription and memoration that character- izes the new technologies? Do they not impose syntheses, and syntheses conceived still more intimately in the soul than any earlier technology has done?[141]

Thus Lyotard asks whether some new, unknown possibility could be opened up by this new technology; or whether, on the contrary, the new technology favours only a synthesis that is ever more efficient and hegemonic, i.e. automation. This question was posed to the philosophers of writing, or of *mnemotechnics*. *Logos* is confronted with the clear mirror, in order to think whether it is possible to realise the clear mirror with *techno-logos*.

As has already been suggested, retrospectively we might ask whether the anamnesis that Lyotard refers to here is not very similar to the emptiness proposed by Nishitani. Indeed, they come from the same tradition, if not the same Zen master. Lyotard seeks to overcome European modernity through an anamnesis which, as he knows, is the ground of East Asian thinking. However, he was probably not aware that this same

---

140. J-F. Lyotard, 'Anamnesis of the Visible', *Theory Culture Society* 21 (2004), 118.
141. Lyotard, '*Logos*', 57 (italics added).

anamnesis was also its greatest weakness when it came to its confrontation with modernisation. Moreover, Lyotard's analysis does not yet touch upon the real problem, which is historical, techno-logical, and geo-political. Lyotard hopes that the 'clear mirror' might negate the tendency toward the totalisation of the system, thus enabling egress from the system as *Enframing*; that it might resist the hegemony of the industrialisation of memory by deviating from its axis of time, which he calls 'common time'.[142] In this sense, he hopes that the postmodern can take up the non-modern and use it as a conceptual tool to overcome the modern. The simple opposition between non-modern and modern has to be problematized, though; and the postmodern, in so far it wants to be a global rather than only a European project, has to re-position itself as an *Aufhebung* that seeks a resolution of the incompatibilities between different ontologies, different *epistemes*.

Let us say a few words about this global axis of time which has become hegemonic through globalisation. I have hinted above that we must move away from the visual image of the globe, since it carries with it the question of inclusion and exclusion. The notion of cosmos as 'house' and sphere originates from an antique European cosmology; a 'stimulating image of an all-encompassing sphere' exemplified by the Ptolemaic model, as Peter Sloterdijk has rightly claimed, which has survived 'until the twentieth century'.[143] In contrast to the image of the globe, Sloterdijk proposes a theory of foam, which he calls a 'polycosmology'. We may be attracted by Sloterdijk's new visual, spatial form of bubbles, which he proposes as the

---

142. Lyotard, '*Logos*', 47.

143. P. Sloterdijk, *In the World Interior of Capital: For a Philosophical Theory of Globalization*, tr. W. Hoban (Cambridge: Polity, 2013), 28.

basis of a 'discrete theory of existence'.[144] However, reviewing Sloterdijk's recent comments on refugee policy may prompt us to ask whether these autonomous bubbles do not conceal an exclusive and seemingly fascist tendency: in an interview with the German political magazine *Cicero* in January 2016, Sloterdijk criticized Angela Merkel's refugee policy, claiming that 'we haven't learned to praise (*Lob*) borders,' and that 'Europeans will sooner or later develop an efficient common border policy. In the long run the territorial imperative prevails. After all, there is no moral obligation to self-destruction.'[145] Do the participations of bubbles only confirm the irreducibility of borders? And doesn't their allure still leave us trapped in the question of territory and inclusion-exclusion?

The real danger of globalisation seems twofold: in consists firstly in a submission to the pure determination of time and of becoming by technologies, as examined above, and secondly in the attempts to overcome modernity, which all too easily turn into fascist and fanatical movements against 'deracinated peoples'. We will conclude the first point here, and will address the second point in the following section.

Towards the end of *Gesture and Speech*, Leroi-Gourhan raised a problem of rhythm that arises with the synchronisation effect of technological systems: 'Individuals today are imbued with and conditioned by a rhythmicity that has reached a stage of almost total mechanicity (as opposed to humanization).'[146] The invitation to move from spatial metaphor to temporal

---

144. P. Sloterdijk, 'Spheres Theory. Talking to Myself About the Poetics of Space', *Harvard Design Magazine* 30 (Spring/Summer 2009), 1–8: 7.

145. P. Sloterdijk, 'Es gibt keine moralische Pflicht zur Selbstzerstörung', *Cicero Magazin für politische Kultur*, 28 January 2016.

146. Leroi-Gourhan, *Gesture and Speech*, 310.

experience is an invitation to rethink rhythms that are in the process of synchronising and becoming homogeneous, following the triumph of the global technological systems that exist in every domain of our daily lives and traverse every territory: telecommunications, logistics, finance, etc. It is this rethinking that must be the main task of a programme of 're-orientation' after Lyotard's postmodern 'disorientation', a programme that aims to go beyond the opposition between the global and the local as the constitution of cultural and political identity. Instead of negating technology and tradition, such a programme will have to open itself to a pluralism of cosmotechnics and a diversity of rhythms by transforming what is already there. The only way to do this is to unmake and remake the categories that we have widely accepted as technics and technology.

In contrast to Lyotard, Eastern attempts to overcome modernity—whether the Kyoto School's fanatical proposal to overcome it by war or Mou Zongsan's optimistic programme of transcending it by descending from the *liangzhi*—failed because they were not able to overcome the time-axis constituted by the technological unconsciousness of modernity on its global scale. Nishitani's strategy was to escape this time-axis by enveloping it in order to give it a new ground—absolute nothingness. Mou's strategy was to descend to this time-axis by contemplating it, in the hope of being able to integrate it, as when he says that 'one heart/mind opens two doors/perspectives (一心開兩門)'. Ultimately what is problematic is that, in both cases, a dualism is presented as a solution. This dualism is not so much a Cartesian one—and indeed both thinkers were very much aware of the problem of Cartesian dualism, and their philosophies also aimed to overcome it—it rather consists in the fact that technics as constitutive of Dasein and *Weltgeschichtlichkeit* is undermined as a mere possibility of the *xin*. All three attempts, we may conclude, are failed ones. However, the way that these questions were raised will allow us to formulate another programme. Lyotard's speculative question has lost none of its power today, because the real question is not whether Chinese or Japanese traditions can give rise to science and technologies, but rather how they can appropriate the global axis of time to radically open up a new realm for themselves, in the way that Lyotard described (but in the opposite direction), and how they can do so without regressing into dualism.

§26. THE DILEMMA OF HOMECOMING

What should we take from these attempts to overcome modernity? Attempts to take a position cleaving to Heidegger's interpretation of philosophy and technology ended up with metaphysical fascism. The Kyoto school's adoption of Hegelian dialectics and Heidegger's mission of philosophy as the theory of the Third Reich to achieve the East Asia Co-prosperity Sphere[147] led not only to a metaphysical mistake but also to an unforgivable crime. However, it is not enough to criticize them simply out of moral indignation: Heidegger did point out a problem that is produced by the planetarization of technology, namely the destruction of tradition and the

---

147.   Hajime Tanabe, in 'On the Logic of Co-prosperity Spheres: Towards a philosophy of Regional Blocs' (1942), presented the project as a Hegelian dialectics which, according to him, will lead to equality of nations; in 1933, Tanabe responded to Heidegger's rectoral address with a series of three articles in a Japanese newspaper 'The Philosophy of Crisis or a Crisis in Philosophy: Reflections on Heidegger's Rectoral Address', in which he argued against Heidegger's prioritisation of the Aristotelian *theorein*, and proposed to consider philosophy as a more active engagement in the political crisis, exemplified by Plato's two visits to Syracuse. The two articles are collected in D. Williams, *Defending Japan's Pacific War: The Kyoto School Philosophers and Post-White Power* (London: Routledge, 2005).

disappearance of any 'home'. But it is a question that must be taken beyond a critique of nationalism, so as to reconsider the grave consequences brought about by technological globalization. A failure to understand this dilemma will end up in the fanaticism of the Kyoto School, which sought to reestablish a world history even at the expense of a total war; or that of Islamic extremism, which believes it can overcome the problem with terror. The cinders of fanaticism will not be extinguished without a direct confrontation of technological globalization, without which it will spread everywhere, both inside and outside Europe, in different forms. The first two decades of the twenty-first century reflect this incapacity to overcome modernity.

The theory of the Russian new right Heideggerian thinker Aleksandr Dugin, meanwhile, can be given as a recent representative example of the tendency to appropriate the 'homecoming' of philosophy as a response against technological planetarisation. Dugin proposes what he calls a 'fourth political theory' as a successor to the major twentieth-century political theories, namely fascism, communism, and liberalism.[148] This new programme is a continuation of the 'conservative revolution' usually associated with Heidegger, Ernst and Friedrich Jünger, Carl Schmitt, Oswald Spengler, Werner Sombart, Othmar Spann, Friedrich Hielscher, Ernst Niekisch, and, more notoriously, Arthur Moeller van den Bruck (1876–1925), whose 1923 book *Das Dritte Reich* considerably influenced the German nationalist movement, which saw modern technology as a great danger for tradition and turned against it. Modernity seems to Dugin an annihilation of tradition, while postmodernity is 'the

---

148. A. Dugin, *The Fourth Political Theory*, tr. M. Sleboda and M. Millerman (London: Arktos Media, 2012).

ultimate oblivion of Being, it is that "midnight", when Nothingness (nihilism) begins to seep from all the cracks'.[149] Dugin's
proposal to overcome both modernity and postmodernity consists in following in the footsteps of Van den Bruck by proposing
that 'conservatives must lead a revolution'.[150] Dugin's idea is go
back to the Russian tradition and to mobilize it as a strategy
against technological modernity. He concretizes this idea in
what he calls the 'Eurasia movement', which is both a political
theory and an *episteme,* in the sense that it uses tradition as an
episteme 'opposed to an unitary *episteme* of Modernity, including science, politics, culture, anthropology'.[151] Even though the
proposed reestablishment of this new *episteme* resonates with
what we have thus far demonstrated, Dugin's programme fails
to develop it further into any philosophical programme, and it
becomes a mere conservative movement.

The 'conservative revolution' is invariably a reactionary
movement against technological modernisation; Heidegger
was one of the first to have transformed this question into a
metaphysical one, namely that of modern technology as the
completion of metaphysics. But Heidegger left open the possibility of a 'homecoming' to the Presocratics. In doing so he may
have been alluding to Hölderlin's lyrical novel *Hyperion,* which
consists of letters between a Greek, his lover, and a German
interlocutor. From the letters, we know that Hyperion once
left his country and travelled to Germany to acquire Apollonian
rationality.[152] However, he found life in Germany unbearable

---

149. Ibid., 22.

150. Ibid., 132.

151. Ibid., 136.

152. J. Young, *The Philosophy of Tragedy From Plato to Žižek* (Cambridge:
Cambridge University Press, 2013), 101.

and went back to Greece, to live as a hermit. Ancient Greece for Hölderlin is an 'experience' and 'knowledge' of a singular historical moment, when technics and nature are presented in tension and conflict.[153] Heidegger appropriated this in his own diagnosis of the contemporary technological situation, and presented it as a 'recommencement'. It is not difficult to see the common ground of the political programmes of Heidegger, the Kyoto School, and Dugin in this notion of a homecoming. The homecoming of philosophy as a recommencement beyond modernity is not only a refusal of technology, characterised by the Heidegger of the 1930s and '40s as 'machination (*Machenschaft*)', a precursor to the term *Gestell*.[154] The renunciation of metaphysics is based on the hope that something more 'authentic' can be revealed—the truth of Being. The truth of Being is however not universal, since it is only revealed to those who went back home, not to those who are not at home, and definitely not to those who stand between the people (*Volk*) and their homecoming. The latter are subsumed under the category of the mass (*das Man*), and of course the Jewish people figure foremost in this category in the *Black Notebooks*, in which what Donatella Di Cesare describes as a 'metaphysical anti-Semitism' prevails: in this reading of history of metaphysics, the Jews become those who have completed and amplified a metaphysical deracination:

> The question of the role of World Jewry [*Weltjudentum*] is not a racial question [*rassisch*], but the metaphysical

---

153. D. J. Schmidt, *On Germans and Other Greeks* (Indianapolis: University of Indiana Press, 2001), 139.

154. I. Farin, 'The Black Notebooks in Their Historical and Political Context', in I. Farin and J. Malpas (eds), *Reading Heidegger's Black Notebooks 1931–1941* (Cambridge, MA: MIT Press, 2016), 301.

question [*metaphysisch*] concerning the kind of humanity [*Men-schentümlichkeit*], which, free from all attachments, can assume the world-historical task of uprooting all beings [*Seiendes*] from Being [*Sein*].[155]

The *Judenfrage* and the *Seinsfrage* constitute an ontological difference, but for Heidegger, *Juden* is not something station-ary like a being-present-at-hand; rather, it is a force that drives the West towards the abyss of Being. Judaism appropriated the modern development of Western metaphysics, and is spreading 'empty rationality' and 'calculating ability'. Judaism walks hand-in-hand with toxic modern metaphysics:

> The reason why Judaism has temporarily increased its power is that Western metaphysics, at least in its modern development, has offered a starting point for the spread of an otherwise empty rationality and calculating ability, which have, consequently, acquired a shelter [*Unterkunft*] in the 'spirit' [*Geist*] without nevertheless being able to grasp, moving from themselves, the hidden ambits-of-decision [*Entscheidungsbezirke*]. The more original and captured-in-their beginning the prospective decisions and questions, the more they remain inaccessible to this 'race'.[156]

---

155. M. Heidegger, *GA 96 Überlegungen XII-XV Schwarze Hefte 1939–1941* (Frankfurt am Main: Klostermann, 2014), 243; cited and translated by D. Di Cesare, 'Heidegger's Metaphysical Anti-Semitism', in *Reading Heidegger's Black Notebooks 1931–1941*, 181.

156. M. Heidegger, *GA 96. Überlegungen XII-XV Schwarze Hefte 1939–1941* (Frankfurt am Main: Klostermann, 2014), 46; cited and translated by Di Cesare, 'Heidegger's Metaphysical Anti-Semitism', 184.

But it is not only the Jews who are portrayed as a malign metaphysical force and an obstacle to accessing the question of Being; Heidegger also has the 'Asiatics' in his sights here, described as 'barbaric, the rootless, the allochthonic'.[157] It is not entirely clear what is meant by 'Asiatic', but it is clear that it carries the general meaning of 'non-European'. On 8 April 1936, at the Hertziana Library of the Kaiser-Wilhelm Institute in Rome, Heidegger gave a lecture entitled 'Europe and the German Philosophy', in which he began by defining the task of European philosophy:

> Our historic Dasein experiences with increasing urgency and clarity that its future is facing a stark either-or: the salvation of Europe, or [alternatively] its own destruction. But the possibility of salvation requires two things:
>
> 1. The shielding [Bewahrung] of European people from the Asiatics [Asiatischen].
>
> 2. The overcoming of its own rootlessness and disintegration.[158]

What is the historical significance of the spread of the empty rationality and calculation that is the destiny of Western metaphysics? It is presented as a crisis, an emergency which European philosophy is not able to deal with, since it is already

---

157. Summarising Heidegger's discussion on the opposition between the early Greek and the Asiatic, Bambach writes: '[Asia] stands as a name for the barbaric, the rootless, the allochthonic—those whose roots are not indigeneous but who come from an/other place. For Heidegger, Asia comes to signify pure alterity, the otherness that threatens the preservation of the homeland'. C. Bambach, *Heidegger's Roots: Nietzsche, National Socialism, and the Greeks* (Ithaca, NY and London: Cornell University Press, 2003), 177.

158. M. Heidegger, 'Europa und die Deutsche Philosophie', cited by L. Ma, *Heidegger on East-West Dialogue: Anticipating the Event* (London: Routledge, 2007), 112; the original text is reproduced in H. H. Ganders (ed.), *Europa und die Philosophie* (Frankfurt am Main: Klostermann, 1993), 31–41.

planetary. The 'Asiatics', whether inside or outside Europe, are considered to be a threat to Europe; however, the Asiatic countries outside Europe were not able to confront techno-logical modernisation either, and the Kyoto School also tried to follow Heidegger in his retreat into the *thinking* of the *Heimattum*. This in turn legitimated a 'metaphysical fascism', in a 'turn' that is common to Heidegger, the Kyoto School, and more recently their Russian fellow conservative .

This reveals the limits of Heidegger's reading of the his-tory of Western metaphysics and the history of technology (as history of nature).[159] However, we must also ask: Why did Heidegger's metaphysical analysis have such a strong reso-nance in the East? Because, once again, what he described is undeniable: namely, the destruction of tradition—for example, when the village loses its traditional form of life and becomes a tourist site.[160] Although it fell outside of his primary concern for the destiny of Europe, Heidegger seems to have suspected that this experience of modernity would be graver outside of Europe than inside—for example when he writes that, if communism comes to power in China, China will become 'free' for technology. After a hundred years of modernisation, the 'homecoming' of all philosophies, whether Chinese, Japanese, Islamic, or African, will be of increasing concern in the twenty-first century because of accelerated dis-orientation. So how can one avoid the fanaticism of total war or terrorism, or the 'conservative revolution'—a metaphysical fascism that claims to be against fascism?

Everyone, every culture, needs a 'home', but it doesn't need to be an exclusive and substantial place. It is the aim of

159. Heidegger, *GA 95*, 133.

160. Ibid., 80.

this book to show that it is not only necessary to seek alternatives, but that it is possible to do so by opening the question of technics not as a universal techno-logy, but as a question of different cosmotechnics. This involves the re-appropriation of the metaphysical categories from inside a culture, as well the adoption of modern technology into it, transforming it.

In comparison to the Communist appropriation of technology as a means of economic and military competition after 1949, the New Confucians took a different approach toward modernisation. They went back to traditional philosophy, fortunately without invoking the same kind of metaphysical fascism; the reason for their failure is historical and philosophical: firstly, since modernisation took place at such astonishing speed, it increasingly left no time for any philosophical reflection whatsoever, especially given that the Chinese philosophical system had perenially failed to identify the category of *Technik* in itself; secondly, the tendency to reconceptualize technology took a rather idealist approach, and therefore became embedded into a cultural programme without having any profound understanding of technology. Cosmotechnics proposes that we reapproach the question of modernity by reinventing the self and technology at the same time, giving priority to the moral and the ethical.

§27. SINOFUTURISM IN THE ANTHROPOCENE

We could have stopped here, since the question concerning technology in China has been almost fully exhibited: firstly, the destruction of the traditional metaphysics and moral cosmology that once governed social and political life; secondly, the attempts to reconstitute a ground proper to their traditions and compatible with Western science and technology, but which only produce contrary effects; and lastly, the deracination

(*Entwurzelung*) that Heidegger anticipated as an imminent danger in Europe, but which has progressed at a much more tremendous pace in Asia. However, we cannot stop here. We must confront the problematic of the 'homecoming' of philosophy, and go beyond it. For it is evidently impossible for the Chinese to totally refuse a science and technology that have indeed effectively become a past that they have never lived, but which has now been passed to them. It is urgent to take further this inquiry concerning the technological condition that is leading to the widespread feeling of the loss of tradition in Asia today; and the only possible response is to propose a new form of the thinking and practice of technologies.

In 1958, during a roundtable discussion, Nishitani described this deracination with enormous grief:

> Religion is impotent in Japan. We don't even have a serious atheism. In Europe, every deviation from tradition has to come to terms with tradition or at least to run up against it. This seems to explain the tendency to interiority or introspection that makes people into thinking people. In Japan [...] ties with tradition have been cut; the burden of having to come to terms with what lies behind us has gone and in its place only a vacuum remains.[161]

The pace of modernisation is probably even faster in China than in Japan, precisely because China was and still is considered to be a country that had 'modernisation without modernity', while Japan is considered to be a country baptised by European modernity. The second half of the twentieth century was full of experiments for China: the Great Leap, the Cultural Revolution, the Four Modernisations (agriculture, industry, national

---

161.   Cited in Heisig, *Philosophers of Nothingness*, 192.

defence, science and technology), the Market Economy.... Subsequently, over the past thirty years, we have witnessed a huge transformation synchronised to a global technological time-axis characterised by speed, innovation, and military competition. As we have seen, and as Nishitani had already observed, the technological system is now totally separated from any moral cosmology: cosmology becomes astronomy, spirit is despised as superstition, and religion becomes the 'opium of the people'. The separation between tradition and modern life that Nishitani worried about has only amplified and intensified, with the gap being enlarged yet further in China under the reforms of the greatest accelerationist of the socialist camp, Deng Xiaoping. As discussed in Part 1, the acceleration led by Deng Xiaoping, on the advice of the thinkers of the 'Dialectics of Nature', squarely placed China on the same technological time-axis as the West. Following this combined acceleration and synchronisation, though, what still lags behind is Chinese thought. The relation between *Dao* and *Qi* has foundered under the new rhythm introduced by the technological system. It is tempting here to repeat Heidegger in saying that 'night is falling'. All one can see is the disappearance of tradition and the superficial marketisation of cultural heritage, whether through the culture industries or tourism. Amidst the economic boom, one also senses that an end is arriving. And this end is going to be realised on a new scene, that of the *Anthropocene*.

The Anthropocene is considered by geologists to be the successor to the Holocene, a geological period which provided a stable earth system for the development of human civilisation. The Anthropocene is regarded as a new era—a new axis of time—in which human activities influence the earth system in previously unimaginable ways. According to the commentators, there is a rough consensus that the Anthropocene started towards the end of the eighteenth century, marked by the invention of James Watt's steam engine, which triggered the industrial revolution. Since then, *homo industrialis* and its technological unconsciousness has become the major force in the transformation of the earth, and the creator of catastrophes,[162] as human beings become elevated to a 'causal explanatory category in the understanding of human history'.[163] In the twentieth century we observed what the geologists called the 'great acceleration', starting from the 1950s, indicated by the economic and military competition during the Cold War, the shift from coal to oil, etc. On the macro-level we have long observed climate change and environmental damage; on the micro-level, geologists have observed that human activities have effectively influenced the geochemical process of the earth. In this conceptualisation of our epoch, geological time and human time are no longer two separate systems.

The recognition of the Anthropocene is the culmination of a technological consciousness in which the human being starts to realise, not only in the intellectual milieu but also in the broader public, the decisive role of technology in the destruction of the biosphere and in the future of humanity: it has been estimated that, without effective mitigation, climate change will bring about the end of the human species within two hundred years.[164] The Anthropocene is closely related to the project of rethinking modernity, since fundamentally the

---

162. I adopt the term *homo industrialis* from M.S. Northcott, *A Political Theology of Climate Change* (Grand Rapids, MI: Eerdmans, 2013), 105.

163. C. Bonneuil, 'The Geological Turn. Narratives of the Anthropocene', in C. Hamilton, F. Gemenne, C. Bonneuil (eds), *The Anthropocene and the Global Environmental Crisis: Rethinking Modernity in a New Epoch* (London and New York: Routledge, 2015), 25.

164. Northcott, *A Political Theology of Climate Change*, 13.

modern ontological interpretations of the cosmos, nature, the world, and humanity are constitutive of what led us into the predicament in which we find ourselves today. The Anthropocene can hardly be distinguished from modernity, since both of them are situated on the same axis of time.

In brief, there are two responses to the potential danger of the Anthropocene: one is geo-engineering, which believes that the earth can be repaired by employing modern technology (e.g. ecological modernism); the other is the appeal for cultural plurality and ontological pluralism. It is the second response that we have tried to engage with in this book. Many efforts have been made to broach this subject, across anthropology, theology, political science, and philosophy—notably, Bruno Latour's 'resetting modernity' project and Philippe Descola's anthropology of nature. The division between culture and nature in modernity is considered by many anthropologists to be one of the major factors in bringing about the Anthropocene. As Montebello claims, contrary to the Ionian cosmologies, which speak the community of all beings whether living, human, or divine, Cartesian dualism makes the human a special kind of being, one that is detached from nature and makes nature its object.[165] It would be too easy to blame Cartesian dualism as a kind of 'original sin', but it would be also ignorant not to see it as a paradigm of the modern project. Modernity began with the *cogito*, with the confidence in consciousness which allows human beings to master the world, to develop a system of knowledge through the self-grounding of the *cogito*, and to set out a programme of development or progress. Theologist Michael Northcott considers it to have been accompanied by

---

165. Montebello, *Métaphysiques cosmomorphes*, 103.

a loss of theological meaning and a politico-theological failure in the West. As he states:

> The dating of the Anthropocene at the outset of the industrial revolution is then indeed the most appropriate from a theological point of view, since it is with the rise of coal, optics and commerce that the sense of co-agency between Christ, Church and cosmos is lost in post-Reformation Europe.[166]

Northcott's observation resonates with Nishitani's, except that, in Europe, although the Industrial Revolution was experienced as a rupture, there was still a certain continuity because this rupture emerged out of an internal dynamic rather than being the result of an irruption of external forces. It is also true that, in recent reflections on the Anthropocene, some prominent intellectuals have proposed a certain reinvention of political theology and cosmology. Thinkers such as Northcott present the Anthropocene as a moment of change, a *kairos* to be seized.[167] Northcott interprets the deep time of the earth discovered by Scottish geologist James Hutton towards the end of the eighteenth century as *chronos*,[168] and sees the Anthropocene as an apocalypse without the intervention of God, a *kairos* which summons human beings to take responsibility for this crisis.

---

166. Northcott, *A Political Theology of Climate Change*, 48.

167. M. Northcott, 'Eschatology in the Anthropocene: from the *Chronos* of Deep Time to the *Kairos* of the Age of Humans', in Hamilton, Gemenne, and Bonneuil (eds.), *The Anthropocene and the Global Environmental Crisis*, 100–112.

168. Hutton was the first to show that the earth has existed for more than 800 million years, in contrast to the biblical belief that the earth is around 6000 years old. Not only did Hutton's discovery challenge the church, it also proposed a theory of 'the system of the earth', considered to be the groundwork of modern geology.

However, among these proposals, there is a common under-estimation of the problem of modernity, as if it is only a 'dis-turbance', a *Störung*. Let us take the example of Latour's 'Resetting Modernity' project. We can understand 'resetting modernity' by way of a metaphor that Latour himself employs:

> What do you do when you are dis-oriented—for instance, when the digital compass of your mobile phone goes wild? You reset it. You might be in a state of mild panic because you have lost your bearings, but still you have to take your time and follow the instruction to recalibrate the compass and let it be reset.[169]

The problem with this metaphor is that modernity is not a malfunctioning machine, but rather one that works *too well* according to the logic embedded in it. Once it is reset, it will restart with the same premises and the same procedure. There is no way in which we can hope that modernity can be reset like pressing a button—or rather, this *kairos* of the modern may be possible for Europe, although I doubt it, but it certainly will not function like this outside of Europe, as I have tried to show by recounting the failures of China and Japan to overcome modernity: the former ended up amplifying modernity, the latter with fanaticism and war. 'Disorientation' does not mean simply that one has lost one's way and doesn't know which direction to choose; it also means the incompat-ibility of temporalities, of histories, of metaphysics: it is rather a 'dis-*orient*-ation'.

In contrast to the appeals to 'return to nature' or to 'reset modernity', what I have tried to propose here is a rediscovery

---

169. B. Latour, 'Let's Touch Base!', in B. Latour (ed.) *Reset Modernity!* (Karlsruhe and Cambridge MA: ZKM/MIT Press, 2016), 21.

of cosmotechnics as both metaphysical and epistemic project. The question that remains to be further formulated is that of the role to be played by modern technologies in this project. It seems to me that this is the fundamental question for overcoming modernity today. It is less about the role of China in the Anthropocene—although we know that China has contributed greatly to its acceleration[170]—than about how China (for example) will reposition itself in relation to the gigantic force of the earth-human time-axis constituted by modern technologies. How is it possible to connect technological consciousness with the cosmotechnics that we have tried to illuminate here? A sinofuturism, as we may call it, is manifesting itself in different domains. However, such a futurism runs in the opposite direction to moral cosmotechnical thinking—ultimately, it is only an acceleration of the European modern project. If we pay attention to what is happening with digitisation in China now, it confirms our view: as Facebook and Youtube arrives, China censors them and builds a Renren or a Youku which look more or less the same; when Uber arrives, China will adopt it and call it Youbu.... As we can understand, there are historical and political reasons for this, yet this is also the moment when such repetition should be suspended, and the question of modernity raised again.

The dream of China after the two Opium Wars, that of 'surpassing the UK and catching up with the USA', seemed to have been realised in certain way in 2015 when it was confirmed that a Chinese company had been contracted to build a nuclear power station at Hinkley Point in the United Kingdom.

---

170. Since 2008 China has been the greatest producer of carbon dioxide, for example. W. Steffen et al., *The Anthropocene: From Global Change to Planetary Stewardship*, AMBIO 40: 7 (2011).

The successful testing of the atomic bomb in 1974 and the hydrogen bomb in 1976 brought China into the first rank of world military powers, but this nuclear programme remained within China's borders. Building a nuclear plant in the UK, however, was symbolically different. In October 2015, Liu Xiaoming, the Chinese ambassador in London, was invited onto the BBC to talk about the power station. Asked whether the UK could also build a nuclear power plant in China, he replied 'Do you have the money first, do you have the technology, do you have expertise? [...] If you have all this, we certainly would want to have co-operation with you, like the French. We have some co-operation with France'.

In February 2016, during the famous annual TV programme of the state-owned CCTV for celebrating the Chinese New Year, the climax came at the moment when five-hundred and forty robots danced onto the stage, with the singer trilling 'rush, rush, rush, dashing to the peak of the world...' while a dozen drones leapt above the stage and shuttled back and forth between the laser lights. This is perhaps the scene that best symbolises the likely future of the Anthropocene in China: robots, drones—symbols of automation, killing, immanent surveillance, and nationalism. One wonders how far the popular imagination has already become detached from the form of life and moral cosmologies that were central to the Chinese tradition. However, what lies behind the scenes—no matter how awkward it is to admit, and no matter how much it might make us lament the loss of tradition—is the fact that China has succeeded in participating in the construction of the axis of time of modernity, and has become one of its major players (and of course this is true not only of China, but also of many other developing countries). This is especially so if one considers the rapid, ongoing modernisation of China and its infrastructural projects in Africa. The 'modern', which was accidental to Chinese culture, is thus not only being amplified within the country itself, but also propagated within the countries of its Third World partners—and in this sense *it is extending European modernity through modern technology* (according to Heidegger, ontotheology).

Thus the question of the Anthropocene is not only that of measures such as reducing pollution, for example, but that of confronting the axis of time which, as Heidegger already observed, is drawing us towards an abyss. This doesn't mean that such ameliorative measures are not important; on the contrary, they are *necessary* but not sufficient. What is more fundamental is the relation between human and the cosmos (between the Heaven and the earth) that defines cultures and natures. As Heidegger predicted, these relations have slowly passed away, yielding to a general understanding of Being as *Bestand*. Capitalism is *the* contemporary cosmotechnics that dominates the planet. Sociologist Jason W. Moore is right to call it a 'world ecology' which ceaselessly exploits natural resources and unpaid labour to sustain its ecology;[171] Economists Shimshon Bichler and Jonathan Nitzan propose that we consider capitalism as a 'mode of power' that orders and reorders power (as the Greek word *kosmeo* itself suggests).[172] Bichler and Nitzan suggest that the evolution of capitalism is not only the evolution of its adoption of modern science and technology; rather they also share the understanding of a cosmic dynamics: for example, between the late nineteenth

---

171.   J. W. Moore, *Capitalism in the Web of Life: Ecology and the Accumulation of Capital* (London: Verso, 2015).

172.   S. Bichler and J. Nitzan, 'Capital as Power: Toward a New Cosmology of Capitalism', *Real-World Economics Review* 61 (2012), 65–84.

century and the early twentieth century, there was a shift from a mechanical mode of power to one that prioritizes uncertainty and relativity.

It is true that we can find some hints in the ancient wisdom as to how the human-cosmos relation might be reconceptualised as a principle of coexistence, governance, and living. For example, in the *Mencius*, there is a famous dialogue between Mencius and King Hui of Liang, in which Mencius denounces war and proposes to the King another way to govern the country by following the 'four seasons' (*sìshí*):

> If the king understands this, there is no reason to expect the people to be more numerous than they are in neighboring states. If the agricultural seasons are not interfered with, there will be more grain than can be eaten. If close-meshed nets are not allowed in the pools and ponds, there will be more fish and turtles than can be eaten. And if axes are allowed in the mountains and forests only in the appropriate seasons, there will be more timber than can be used. When grain, fish, and turtles are more than can be eaten, and timber is more than can be used, this will mean that the people can nourish their lives, bury their dead, and be without rancor. Making it possible for them to nourish their lives, bury their dead, and be without rancor is the beginning of kingly government.[173]

A similar comment on governing according to the season is found in the writings of Mencius's contemporary Xunzi (荀子, 313–238 BC).[174] The ancient Chinese wisdom has been

---

173. Mencius, *Mencius*, tr. I. Bloom (New York: Columbia University Press, 2009), Book I, 3.

174. In contrast to Mencius's proposal that human nature is good, Xunzi

ceaselessly repeated in the past decades in view of the ecological crisis and rampant industrialization, and yet what we
have heard are only constant catastrophes. The *Li* (rites)
have become purely formal, to the point where, ridiculously,
one prays to the Heaven so that one can exploit more the
earth in order to profit more. It is not that there is no awareness of the problems, but rather that pragmatic reason—the
reason that seeks to adapt itself so as to profit from globalization—prevents us from raising the deeper questions of
cosmotechnics and *episteme*. The cosmotechnical relation to
the cosmos—not only as intimacy but also as constraint, is in
most cases disregarded in the industrial mode of production.
Huge varieties of knowledge and knowhow are replaced by
the domination of a global *episteme* imposed by capitalism.
This technological becoming of the world has to be challenged
in order to interrupt its hegemonic synchronization and to
produce another mode of coexistence. However, although
we do not resign ourselves to admitting that the principles of
Chinese philosophy are simply rendered obsolete or anach-
ronistic by the progress of the global time-axis, neither will
this question be answered by a superficial espousal of the
'spiritual', or by inscribing technology within a 'philosophy of
nature' imagined to emanate from ancient lore, and supplying
a pacifying metaphysics that simply mitigates the disquiet created by dis-orientation (i.e. the model of Zen or Dao as 'self-
improvement' for the consumer). Reappropriating technology

---

insists that it is evil, and is the reason why education is important. Xunzi is in
agreement with Mencius on the question of ecology, however: the King, as
sage, should impose laws to protect natural resources when they are in the
process of growing, for example when plants and trees are pushing in the
forest, axes should be forbidden to enter the forest ('草木榮華滋碩之時，則
斧斤不入山林，不夭其生，不絕其長也').

complicates the project of 'overcoming modernity', since this can only be a global project—one that is constituted by, and struggles against, a common time-axis; any retreat from the question of the global will not give us a better solution than slow disintegration. Therefore, world history must be approached from this standpoint.

§28. FOR ANOTHER WORLD HISTORY

By positing this common time-axis and world history, are we, as postcolonial scholars argue, trapped in a sort of historicism, accepting a certain narrative of European modernity as the pivot of world history?[175] This question certainly deserves our attention, since it can be dangerous to dress up an old problem in new clothes. Yet this is not merely a question of narratives, but rather of a technical reality that cannot be reduced to the level of discourses alone. One of the dangers of arguing that world history is merely a narrative, and that therefore it is possible to find an exit from it via another narrative, is that it ignores the materiality of such world history, and takes the relation between technics and thinking, between *Dao* and *Qi*, to be a matter of texts alone. We know, for example, that the historicism that developed between roughly 1880 and 1930 among German historians and Neo-Kantians was bankrupt after the World Wars;[176] it is not history as a narrative that is our problem, but rather how it functions in material terms. A new constitution of time, and therefore of a new world history, I would claim, must consist not simply in a new narrative, but rather in a new *practice* and *knowledge*, one that is no longer

---

175.   D. Chakrabarty, *Provincializing Europe: Postcolonial Thought and Historical Difference* (Princeton, NJ and Oxford: Princeton University Press, 2000).

176.   See Bambach, *Heidegger, Dilthey, and the Crisis of Historicism*.

totalised by the time-axis of modernity. This is a difference of position in regard to postcolonial critiques that must be emphasised.

In this spirit, let us briefly examine some of the ideas of the postcolonial historian and scholar Dipesh Chakrabarty, as presented in his wonderful and provocative *Provincializing Europe*, a book dedicated to a thorough critique of historicism and the notion of Europe as the axis of the historical narrative of modernity. Chakrabarty uses Heidegger to problematise Marx's concept of history as a paradigm of 'History 1 vs History 2', using the contrast between the ready-to-hand (*zuhanden*) and the present-at-hand (*vorhanden*):

> Heidegger does not minimize the importance of objectifying relationships (History 1 would belong here)—in his translator's prose, they are called 'present-at-hand'—but in a properly Hei- deggerian framework of understanding, both the present-at- hand and the ready-to-hand retain their importance; one does not gain epistemological primacy over the other. History 2 cannot sublate itself into History 1.[177]

A few pages later, Chakrabarty states more clearly what he means by History 1 and 2: when capital as a philosophical- historical category is analysed as the transition of History 1 through translations, it becomes a universal and empty abstraction; however, History 2 is what opens the 'historical difference' and therefore involves a different kind of transla- tion, one that is constituted by an irreducible *difference*. In this

---

177.   Chakrabarty, *Provincializing Europe*, 68.

sense, the Heideggerian ready-to-hand can be mobilised to resist the 'epistemological primacy' of History 1:[178]

> History 1 is just that, analytical history. But the idea of History 2 beckons us to more *affective* narratives of human belonging where life forms, although porous to one another, do not seem exchangeable through a third term of equivalence such as abstract labor.[179]

The problem of this whole analysis is the unexplained *Zuhandene*. As I have shown elsewhere, *Zuhandene* are fundamentally technical objects in our everyday life. They are not *Vorhandene*, objects (*Gegenstand*) that stand (*stehen*) over against (*gegen*) the subject. The temporality of the *Zuhandene* is defined by equipmentality (*Zeuglichkeit*). For example, when we use a hammer, we do not need to thematise it; rather, we use it as if we know it already. Heidegger's *Zuhandenheit* ('readiness-to-hand') is a composite of discursive and existential relations which constitute the temporal dynamic of technical objects, but also of technical systems.[180] We live in a world composed of more and more technical

---

178. Ibid., 239

179. Ibid., 71.

180. See my *On the Existence of Digital Objects*, chapter 3, where I propose an ontology of relations to describe the dynamic between what I call 'discursive' and 'existential' relations. These two types of relations should not be confused with what, in mediaeval philosophy, are known as *relationes secundum dici* and *relationes secundum esse*, since the latter still retain the notion of substance, which relational ontology seeks to move away from. In short, *discursive* relations are those that can be said and hence materialised in different forms, not excluding causal relations—for example, drawings, writings, physical contacts of pulley and belt, electrical currents, and data connections; *existential* relations are relations to the world that are constantly modified by the concretisation of discursive relations.

objects developed during different periods of history, pos-
sessing different temporalities; and the opposition between
Historie and *Geschichte*, *present-at-hand* and *ready-to-hand*
as fundamental categories is not sufficient to account for
historicity itself. This was precisely the point at which we
were able to stage a confrontation between Stiegler's and
Nishitani's interpretation of world historicity. Chakrabarty's
characterisation of *Zuhandenheit* as the life-world is an intuitive and indeed very interesting way to conceptualise alternative histories against the history of colonisation, since the
ready-to-hand resists any reduction to essence; however, it
is not possible to deduce a historical conception based on
the *Zuhandene* without recognising their nature as technical
objects, and the fact that they cannot exist alone as technical
objects, but only in a world—a world which is increasingly
becoming a unified and globalised system.

As Chakrabarty says, the axis of time that synchronises
global activities is becoming more powerful and at the same
time more homogeneous; this is precisely what we call 'modernisation'. However, I do not agree that one can reduce this
axis of time simply to narrative, and thereby easily 'provincialize'
it. Chakrabarty's critique exemplifies the problem of many
postcolonial theories, which tend to reduce political and material questions to the register of inter-textuality in comparative
literature. Modernity qua technological unconsciousness will
necessarily continue to propagate in other cultures and civi-
lisations. The declaration of the end of modernity in Europe
does not mean that modernity in general ends, since it is only
in Europe that such a technological consciousness is seized
both as a fate and as a new possibility (as in Nietzsche's nihilism). It is a *necessity* for other cultures because technological
unconsciousness is promoted by global military and economic

competition, so that technological modernisation becomes inevitable. It was under such conditions that China found it necessary to speed up its technological development—constant tensions with the Soviet Union and the USA during the Cold War, and then the arrival of the market economy, only pushed it to exhaust all natural and human resources in order to maintain a constant growth in GDP. So the question is not simply that of developing new narratives, or of looking at world history from the point of view of Asia or of Europe, but rather that of confronting this time-axis in order to overcome modernity *through* modernity, meaning through the reappropriation of modern technologies and technological consciousness.

The sort of cosmopolitanism constituted by global commerce as *cosmopolitan right*, as envisioned by Kant in his *Perpetual Peace* (1795), as well as in his projection of a common becoming in his *Idea for a Universal History from a Cosmopolitan Point of View* (1784), has been to some extent realised with the various technologies of reticulation in force today (e.g. different forms of networks, transportation, telecommunication, finance, anti-terrorism, etc.). One might argue, as does Jürgen Habermas, that the kind of reason that Kant described has not yet arrived, that the project of Enlightenment is not yet complete. However, the question seems to be no longer about completing a universal reason in the Kantian and/or Hegelian sense, but rather about reconstituting a variety of cosmotechnics able to resist the global time-axis that has been constructed by modernity. Having criticized the European colonists and traders, Kant observes that China and Japan have wisely decided upon their policy against these foreign visitors: the former allow contact, but no entry into the territory; the latter limit their contact with the Dutch while at the same time treating them as criminals.[181] But such 'wisdom' has proved impossible in the context of globalisation; and it is also impossible to go back to this state of isolation—for what was external (e.g. trade) is now internal to the country (e.g. through financial and other networks).

But today the task of overcoming modernity through modernity inevitably brings us to the question of specificity and locality. Locality is not the reassuring alternative to globalization, but its 'universal product'.[182] If we want to talk about locality again, then we must recognize that it is no longer an isolated locality— the self-isolated Japan or China, disconnected or remote from the global time-axis—but must be a locality that appropriates the global instead of being simply produced and reproduced by the global. The locality that is able to resist the global axis of time is one capable of confronting it by radically and self-consciously transforming it—rather than merely adding aesthetic value to it. The local cannot stand as an opposition to the global, otherwise it will risk defaulting to some kind of 'conservative revolution', or even facilitating metaphysical fascism. I have attempted here to take a first step towards deviating from the conventional reading of Chinese philosophy as a mere moral philosophy, to reassess it as cosmotechnics, and to put forward the traditional metaphysical categories as our contemporaries; I have also aimed to open up the concept of technics as multi-cosmotechnics, consisting of different irreducible metaphysical categories. The reappropriation of modern technology from the standpoint of cosmotechnics

---

181. I. Kant, 'Toward Perpetual Peace', tr. D.L. Colclasure, in *Toward Perpetual Peace and Other Writings on Politics, Peace, and History* (New Haven, CT and London: Yale University Press, 1996), 83.

182. Invisible Committee, *To Our Friends* (Cambridge MA: Semiotext(e), 2014), 188–9.

demands two steps: firstly, as attempted here, it demands that we reconfigure fundamental metaphysical categories such as *Qi-Dao* as a ground; secondly, that we reconstruct upon this ground an *episteme* which will in turn condition technical invention, development, innovation, in order that the latter should no longer be mere imitations or repetitions.

In speaking of China or East Asia in general, the question—central to our thesis here—is how the *Qi-Dao* relations that we sketched out in Part 1 might contribute to the discussion on diversity or pluralism. In outlining the lineage of the *Qi-Dao* relation, we do not intend to suggest a return to an 'original' or 'authentic' relation between *Qi* and *Dao*, but rather to forcefully open up a new understanding of *Dao* in relation to the global axis of time. If we look for examples in the past, the emergence of different schools (including Confucianism, Daoism, etc.), Neo-Confucianism, and New Confucianism were invariably responses to political crisis or to the decline of spirit. Each attempted to renew an *episteme* based on the reinterpretation of the tradition by way of metaphysical categories. This *episteme* would in turn condition political, aesthetic, social, and spiritual life (or form of life) and serve as a force of creation and constraint upon knowing. For example, in the tea ceremony or calligraphy, where the use of *Qi* is no longer one that aims at a certain end, but rather one that aims for a totally different experience. In these instances *Qi* is transformed into a higher purpose, which we may call, following Kant, 'purposiveness without purpose'. These forms of aesthetic practice have been widely carried out in China from ancient times up to the present day. Owing to the modernization of everyday life they are becoming less widespread, even if some of them are now being revived in the context of the marketing strategies of our consumer society. The question is not simply that of retreating into aesthetic experience, but rather that of refining the philosophical thinking that may be contained within it. What is central to such a proposal to trace and seek a philosophy of technology in China is to systematically reflect on the relation between technics, and the unity of the cosmic and moral order—which will allow us to begin to reflect once more on the production and use of technology.

There remain many questions to be further reflected upon and concretely experimented with: How can such a form of experience be imagined in relation to information technologies—computers, smartphones, robots, and so on? How can we talk about *Qi-Dao* in relation to diodes, triodes, and transistors, examples that Gilbert Simondon used to discuss the mode of existence of technical objects? How are we to renew the relation with non-humans after a hundred years of modernisation? Technological development has overflowed the framework of the ancient cosmotechnics, to the extent that those ancient teachings such as Daoism, Buddhism, or even Stoicism become dogmas and are consequently adopted as no more than self-help methods; in the 'best' cases they are transformed into something like the 'Californian ideology'.[183] However, we maintain that it is possible to raise these questions anew, and to approach them from the cosmotechnical point of view—and not that of *Gelassenheit*—according to different orders of magnitude, from that of the cosmos to that of *ch'i*. Simondon's analysis of the TV antenna, discussed above (§2), seems to me a good example of how we might imagine the compatibility between cosmotechnical thinking and modern technology.

---

183. The hippie movement in the United States during the 1960s, where a rather westernised form of Zen Buddhism was adopted as the religion of many hackers.

The concept of cosmotechnics—beyond cosmologies— therefore hopes to reopen both the question and the multiple histories of technology. In other words, in using China as an example, and proposing to take up the *Qi-Dao* cosmotechnics as the ground and constraint for the appropriation of modern technology, we aim to renew a form of life and a cosmotechnics that would consciously subtract itself from and deviate from the homogeneous becoming of the technological world. This is impossible without a reinterpretation of our tradition and its trans*formation* into a new *episteme*. And it will also involve another form of translation: no longer a translation based on *equivalence*, for example from metaphysics into *xing er shang xue* or *techne* into *jishu*, but a translation based on *difference*, a *translation* that allows a *transduction* to take place.

Transduction, as understood by Simondon, implies the progressive structural transformation of a system triggered by incoming information—part of the individuation of civilisation, in which progress is characterised by 'internal resonances'. In an article entitled 'The Limits of Human Progress: A Critical Study',[184] a response to Raymond Ruyer's 1958 article of the same title on the question of technological acceleration in relation to the limits of human progress, Simondon proposed to consider the physical concretisation of technical objects as a limit on civilisation. Ruyer had rejected Antoine Cournot's idea that technological progress was a regular and linear accretion, describing it rather as an 'accelerated explosion', and argued that the exponential acceleration of technology will stop at some point.[185] We cannot elaborate on Ruyer's arguments

---

184. G. Simondon, 'The Limits of Human Progress: A Critical Study' (1959), *Cultural Politics* 6:2 (2010), 229–36.

185. R. Ruyer, 'Les Limites du progrès humain', *Revue de Métaphysique et de Morale* 63: 4 (1958), 412–27: 416.

here, but it is interesting to note that, by the end of the article,
he states that, even though the Industrial Revolution in the
eighteenth and early nineteenth centuries brought misery to
a large part of the population, 'once the technical skeleton is
stabilised, life can begin its games and fancies anew'.[186] Ruyer's
argument may find resonance with the pragmatists in China:
let development take its course, and please bear with the
catastrophes—we will repair 'nature' afterwards. Simondon,
rather than presupposing a definite end to human progress,
proposes to understand human progress in terms of cycles
characterised by the internal resonance between human being
and objective concretisation:

> [W]e can say that there is human progress only if, when passing
> from one self-limiting cycle to the next, man increases the part of
> himself which is engaged in the system he forms with the objec-
> tive concretisation. There is progress if the system man-religion
> is endowed with more internal resonance than the system man-
> language, and if the man-technology system is endowed with
> a greater internal resonance than the system man-religion.[187]

Here Simondon identifies three cycles, namely 'man-language',
'man-religion', and 'man-technology'. In the 'man-technology'
cycle, Simondon observes a new objective concretisation,
which is no longer that of natural language or religious rituals,
but that of the production of 'technical individuals'. It is possible that technical concretisation may not produce any internal
resonance, and therefore may not lead to a new cycle. This,
we might say, constitutes Simondon's critique of modernity,

187. Simondon, 'The Limits of Human Progress', 231.

FOR ANOTHER WORLD HISTORY

a critique that finds its concrete example in today's China as well as in most parts of Asia, where one finds a entropic becoming driven by capitalism (the dominant cosmotechnics) leading nowhere,[188] and with no resonance—the universalisation of naturalism in Descola's sense. This is the danger posed to all of us in the Anthropocene. Here, producing an internal resonance is the task of translation. The 'internal resonance' we seek here is the unification of the metaphysical categories of *Qi* and *Dao*, which must be endowed with new meanings and forces proper to our epoch. One will certainly have to understand science and technology in order to be able to transform them; but after more than a century of 'modernisation', now is the moment to seek a new form of practice, not only in China but also in other cultures. For China as addressed in this book is only one example, and only one of many possibilities. This is where imagination should take off and concentrate its efforts. The aim of this book has been to put forward such a new translation based on difference. It is only with this difference, and with the capability and the imagination to assert this difference in material terms, that we can stake a claim to another world history.

---

188. I relate the term 'entropic' to what Lévi-Strauss in *Tristes Tropiques* called 'entropology', a term he suggests to rename his own discipline, anthropology, which describes the disintegration of cultures under assault from Western expansion: 'anthropology could with advantage be changed into "entropology", as the name of the discipline concerned with the study of the highest manifestations of this process of disintegration'. C. Lévi-Strauss, *Tristes Tropiques*, tr. J. Weightman and D. Weightman (New York: Penguin Books, 1992), 414. The term has recently been invoked by Bernard Stiegler, when he called the anthropocene an 'entropocene', in the sense that it constantly produces *hubris*, see B. Stiegler, *Dans la disruption: comment ne pas devenir fou* (Paris: Editions les Liens qui Libèrent, 2016); Jason W. Moore calls it the Capitaloscene, in the sense that the anthropocene is fundamentally a stage of the world ecology of capitalism (Moore, *Capitalism in the Web of Life*).

INDEX OF NAMES

INDEX OF SUBJECTS